T0254561

SpringerBriefs in History of Science and Technology

More information about this series at http://www.springer.com/series/10085

Alexander Pechenkin

The History of Research on Chemical Periodic Processes

 Springer

Alexander Pechenkin
Department of Philosophy
Lomonosov Moscow State University
Moscow, Russia

ISSN 2211-4564   ISSN 2211-4572   (electronic)
SpringerBriefs in History of Science and Technology
ISBN 978-3-319-95107-2   ISBN 978-3-319-95108-9   (eBook)
https://doi.org/10.1007/978-3-319-95108-9

Library of Congress Control Number: 2018946672

Printed on acid-free paper

This Springer imprint is published by the registered company Springer Nature Switzerland AG
The registered company address is: Gewerbestrasse 11, 6330 Cham, Switzerland

Preface

"All men by nature desire to know. An indication of this is the delight we take in our senses; for even apart from their usefulness they are loved for themselves".[1]

As I was working on this manuscript, I kept reciting this Aristotle's quote from the very beginning of his «Metaphysics». My inquiries, presented in this manuscript, did not aim to contribute to the progress of modern-day technology or to proliferate useful innovations. My exploits were driven primarily by my curiosity and an aspiration to trace the life of past scholars and to reconstruct their trains of thought that led to their spectacular work. As a rule, the scholars whose scholarly life I lay out in this manuscript did not enjoy the personal wealth or unusual funding benefits. Prestigious awards did not single out their accomplishments as being of particular significance to their contemporaries. Moreover, their most groundbreaking research was frequently treated as subsidiary to their mundane tasks for which they earned their regular salaries and had to be conducted at the time of leisure or after hours.

Although several scholars involved in investigating chemical periodicity became recognized as prominent scholars during their lifetime, this fame did not stem from their contributions to the field of chemical periodicity. Instead, their academic recognition was initially prompted by their routine work, which was officially sanctioned as being important at the time based on the needs of industrial development. The importance of the periodicity research only started to be recognized in 1960s, when A. M. Zhabotinsky and his colleagues started their research on Belousov's reaction. Zhabotinsky turned his Ph.D. and Dr. Sci. degrees for his contribution to understanding chemical periodicity. At the time, special conferences on chemical periodicity started to be held in Pushino-na-Oke, a small research town situated in Moscow suburbs.

In 1981, for their contribution for chemical periodicity, A. Zhabotinsky, B. Belousov (posthumously), and several of their colleagues received the Lenin Prize, which at the time was the highest prize in the Soviet Union. In 1987, Ilia

---

[1] Aristotle. Metaphysics. A. 980a.

Prigogine, a Belgian physicist and chemist, received the Noble Prize for his contribution to the concept of the dissipative structure. The "shared example" (using T. Kuhn's terms) of the dissipative structure is the Belousov–Zhabotinsky reaction.

Through this manuscript, I explore whether this 1960–1980 success in studying chemical periodicity would be possible without the hard and enthusiastic work of dozens researchers in the field of chemical periodicity and to what extend the success of their exploits was grounded in the scholarship of the last decades of nineteenth century. Both the Aristotle's statement and the history described in this book bear an optimistic promise. This optimism is particularly important for the scholars of present-day Russia. The great scholarship tradition of the former Soviet Union has been anything but completely annihilated in the aftermath of the collapse. Russian scientists of today survive on minimal wages and meager funding, suffering perpetual bureaucratic injustice and popular scorn. The history presented in this book, nevertheless, gives us hope that there is a natural quest for knowledge, the quest which can overcome all hardships.

Moscow, Russia   Alexander Pechenkin

Acknowledgements

Chemical Heritage Foundation supported my work by supplying library fellowships to allow me to use the resources of Beckman library (2004, 2005, 2006, and 2017) and by Robert W. Allington fellowship (2011).

I also wish to thank Prof. Joseph E. Earley (Georgetown University) who was kind enough to help me improve my text (Chap. 5, Sects. 5.2–5.4) for publication. Professor Rein Vihalemm comments also were useful. Professor Vihalemm's (1938–2016) worked for University of Tartu (Estonia).

This book uses archival material from the Archives of Medical and Biological Extreme Problems and the Russian State Military Archives and is based on interviews conducted by me with Belousov's colleagues. These colleagues are Profs. Boris B. Moroz, Lev M. Rozhdesvenskii (both interviews conducted on March 3, 2004), and Dr. Gennadi I. Shaposhnikov (interviewed on September 21, 2005). At the time of the interviews, they all worked for the Institute of Biophysics (belonging to the Federal Agency of Public Health), where Belousov was employed. The explanations provided by the university research teacher of A. M. Zhabotinsky, Prof. Simon. E. Shnol', also helped me. Now Shnol' is the only witness of the first steps of the department of biophysics where Zhabotinsky started as a scientist.

The material from Archives of Lomonosov Moscow State University (Fond of the Institute of Chemistry and Chemistry Faculty, Fonds N 9, 14) was used in Chap. 2.

The author also consulted Prof. Vladimir V. Kuznetsov, the head of the Department of Analytical Chemistry at the Mendeleev Moscow Institute of Chemistry and Technology.

The explanations which Prof. Richard Field, one of the creators of the FKN model of the Belousov-Zhabotinsky reaction, provided for the present author cannot be overestimated. In the book R. Field's comments are used in Chaps. 2, 3 and 5.

Professor Mazo, who observed the early history of the FKN model of the BZ reaction, kindly wrote his recollections which are housed in Chap. 4.

The author discussed some historical problems with Prof. Vasilii Vavilin, who directly cooperated with Zhabotinsky. Now Vavilin works for Institute of Water Problems.

Professor Yu. A. Zolotov (Moscow State University, Chemical Faculty, Chair) explained some detailed concerning oscillatory reactions to the present author and helped him to understand some historical details concerning Mikhalev and Schemjakin. Thank you very much.

Doctor Elena Baum-Zaitseva kindly read some fragments of the manuscript and made some comments concerning the text.

Contents

Introduction

"The modern history of research in chemical oscillators started in the USSR in 1951 when B. P. Belousov discovered the concentration oscillations between oxidized and reduced forms of cerium in the process of the oxidation of citric acid by bromate. A. M. Zhabotinsky continued Belousov's research to define a class of the oscillatory reactions of organic compounds oxidation by bromate in the presence of metal ion catalyzer that are now referred to as the Belousov-Zhabotinsky reactions (BZ reactions)" (Field and Burger 1985, p. V). Indeed, the discovery of Belousov–Zhabotinsky reaction led to intensive studies of mechanisms of chemical oscillations developing mathematic apparatus to study kinetics and thermodynamics of these peculiar reactions, and eventually led to the development of nonlinear thermodynamics and new applications of the theory of dynamical systems. Ultimately, the studies of the BZ reaction have been instrumental for conceptual development of mathematical approaches and specifically mathematical modeling in modern chemistry. The BZ reaction shows the traveling waves regime, and the concept of dynamical chaos is becoming actual with respect to the BZ systems (the Belousov 1951 article shows that he already observed traveling waves in his system).

The BZ discovery had influenced the studies of oscillations in biological systems. For example, Dr. D. S. Chernyavsky in his interview (14.5.2005) said to the present author that his study regarding oscillations in photosynthesis that was rejected by journals had been published in 1960 (Chernavskaya and Chernavsky 1960) due to the protection of I. E. Tamm, the great figure in the Soviet Academy of Sciences (Tamm received Nobel Prize in 1958). However, after work by Zhabotinsky's developed interest in chemical oscillations. Soviet journals became to publish such papers without any protection.

True, the real influence of the BZ discovery on biology was not immediate. Oscillation in biological systems has been studied and well-known long before Belousov. However, the BZ reaction led to the development of new conceptual means to analyze these oscillations (see, e.g., Strogatz 1994, p. 4, 245, 255).

With respect to the history of science the BZ reaction caused an unexpected after effect. Chemical phenomena, which were neglected in the historical analysis, were suddenly rediscovered and placed into a new conceptual framework. Indeed, we suddenly realized that William Bray who discovered perhaps the first homogeneous oscillatory reaction in 1921—the conversion of hydrogen peroxide catalyzed by iodate—was influenced but influenced by mathematician, physico-chemist and one of the first bio-mathematicians A. Lotka who predicted chemical oscillations in homogeneous systems. The BZ reaction in many ways changed the predominant chemical "culture". Following the period of initial resistance and outright denial, the oscillatory problematics has become interesting and fascinating. Moving away from a purely mechanistic point of view, chemists became aware of such unexpected concepts as system self-organization, synergetics and formation of structure, chaos and stability of chemical systems. The philosophers who were interested in discussing the world problems suddenly noticed the new field that before escaped their attention. The BZ reaction is the main topic of several historical essays. One of them was written by Zhabotinsky himself (1991, p. 379–386). S. R. Scott's essay which opens his 1994 book is close to the one written by Zhabotinsky. The Zhabotinsky-Scott historical paradigm emphasizes the early experiments and theories concerning oscillation in homogeneous chemical systems, the "Dark age" for oscillatory problematics in chemistry (from the 1920s until the mid-60s) and the fundamental turn to the oscillatory problematics at the 1960s.

However, the other group of historical studies appeared in the 1980s–1990s. These works pointed to the important developments during these "Dark ages" including studies of periodical processes in electrochemistry, in colloid chemistry, and in technology that were important from conceptual point of view (Volter 1985, 1988; Fedorov 1998).

A special chapter of the history of chemical oscillations consists of the literature on Liesegang rings which were discovered in 1896 and sprouted many theoretical and experimental studies by chemists and physicists alike. Although there is no systematic history of research related to Lisegang rings, several books and articles contain historical observations concerning Liesegang's biography, his contribution to the development of chemistry.

Apart from the history of Liesegang rings, there is the literature about the studies on the periodical processes on solution of metals in acids, on electrochemical periodical processes.

We will present a study of the history of chemical oscillations. The author will cover the early history which usually is essentially ignored in many historical essays. The importance of this period cannot be underestimated as the very concept of chemical oscillations took shape in these early studies. Here we shall consider Wilhelm Ostwald's work while avoiding belaboring on the topics that have been already covered in the historical literature.

The next stage: the first books on chemical oscillations. They were published in the "leading chemical" countries: Germany, GB, France, and the USSR (1913–1938). The very concept of chemical periodical processes has been shaped due to these books. We have come to the central points. A special chapter will be dedicated to the biography of Belousov and his research. We will provide an extended reconstruction as the archives contain only a very abridged version of Belousov's autobiography. We shall follow the development of ideas and conceptual framework, which led to the discovery of his famous reaction.

Zhabotinsky's biography and his creative work will be presented in connection with the ideological and organizational development of the Soviet science in 1960s–1970s. The establishing of the Department of biophysics at the School of Physics at the Lomonosov Moscow State University, the popularity of Andronov's school of nonlinear oscillations, the success of applied mathematics in the USSR, the development of scientific institutions in Pushino-na-Oke, in a small town which located nearby Moscow, will be described in this chapter of the book. Because the history of Prigogine's nonlinear thermodynamics has been extensively covered in previous works, we shall concentrate on the structure of this theory. This excursion will allow us to elucidate the influence of the BZ reaction on Prigogine and his co-workers, to describe the conceptual meaning of "dissipative structure" and to follow its implication for the theory of dynamic systems and theoretical biology.

Chapter 6 will be dedicated to the "American line" in the history of chemical oscillations. This section will focus on Bray's research and on the connection of the Bray reaction to the BZ reaction. This section will also consider R. Noyes' and his co-workers' 1950–70s research in chemical kinetics, the 1972–1974 FKN mechanism of the BZ reaction and the 1974 mathematical model named Oregonator.

We will trace not only the history of ideas, but the history of people. People who conducted the oscillatory research in chemistry were mainly enthusiasts, volunteer researchers, who studied chemical oscillators in parallel with their main job. In the 1960s, the situation changed: The research of periodical chemical processes had been transformed to what T. Kuhn called the "normal science."

Some general trends in the development of ideas will be presented including the development of general science on oscillations (Schwingungslehre), the development of nonlinear approach in the theory of oscillations, the transformation of thermodynamics. In this work, I am extensively relying on the original journal publications and to use unique archival material and the author's interview of Russian and American scientists who were involved in the developing of the field.

In their historical essays, Zhabotinsky and Scott consider Robert Boyle's note as a starting point of studies in chemical oscillations (the end of seventeenth century). "Boyle noted a periodic flaring up phosphorous in a loosely stoppered flask" (Scott 1991, p. 3). In this book, we start with P. Munk's and G. Fechner's writings on periodical processes. These writings opened a rather continuous line of publications on periodical chemical processes.

Scott's book contains a table which lists the main discoveries of chemical periodical processes (XIX century and the beginning of XX). In what follows we shall describe some of them by taking historical circumstances into account.

References

Chernavskaya NM, Chernavsky DS (1960) Periodicheskie iavlenia v fotosinteze (The periodical phenomena in photosynthesis). Uspekhi fizicheskikh nauk 72:627–649

Fedorov PP (1998) The history with oscillations. Khimia i zhisn 2:34–35

Field R, Burger M (1985) Oscillations and traveling waves in chemical systems. Wiley, USA.

Scott SK (1991) Oscillations and waves in chemical kinetics, vol 90. Oxford Science Publications, Oxford, NY, Toronto

Strogatz SH (1994) Nonlinear dynamics and chaos. Perseus books, Reading, MA, pp 467

Volter BV (1985) "Kto otkryl kolebatelnye khimicheskie reaktsii (Who discovered the oscillatory chemical reactions?)". Chimia i zhizn 2:8

Volter BV (1988) 'Legenda and byl' o khimicheskikh kolebaniakh (the chemical oscillations: myth and reality)'. Znanie-sila 4:33–37. See also SP Kurdiumov' sait. http://spkurdyumov.ru/introduction/legenda-i-byl/ (in Russian)

Zhabotinsky A (1991) A history of chemical oscillations and waves. Chaos 1:379

Chapter 1
Early Observations

Abstract The first chapter refers to several examples of chemical periodic processes described in XIX. Peter Munk (1804–1860) observed the regular flashes in the process of oxidation of phosphorous. Gustav Fechner (1801–1887) observed fluctuation of an iron electrode immersed in a weakly acid solution of silver nitrate. However, Fechner did not include the problem of chemical periodicity into his text books. Georg Quincke (1834–1924) noted the periodic spreading and accumulation of oil drops on a large water-air surface. He also wrote about the phenomenon of periodic precipitation. The special story is concerned with rhythmic structures obtained by Friedlieb Runge (1794–1887) and published as two books by him. Here chemical periodicity turned out to be connected with art and aesthetics. There is considerable historical literature about Runge and his creative work. In particular, F. Runge is mentioned as one of the predecessors of paper chromatography.

Keywords Oxidation of phosphorous · Oxidation-reduction reaction
Electrochemistry · Electrode · Fluctuation · Precipitation · Sediments
Paper chromatography · Art · Living force

1.1 Peter Munck

In the papers on the oscillation of flame published in the mid of last century, there are reference on Peter Munck's paper which appeared in 1834 (e.g., Frank-Kamenetskii 1939, p. 67; Gray P., Scott, 1985, p. 556; Scott 1991, p. 3). Peter Munck wrote that he conducted experiments with phosphorus: "A small vessel containing phosphorus was used as a lighter for a while. As this vessel was not properly closed phosphorous which had partly been oxidized drew water and stopped to produce light. I let my vessel standing aside without any disturbance. However I accidently noticed that this vessel gave a rather intensive light and I soon perceived that the flashes take place regularly every seventh second" (Munck, p. 216).

Further, Munck described the behavior of his apparatus in different conditions. He discussed the influence of temperature on the periodicity of flashes.

© The Author(s) 2018
A. Pechenkin, *The History of Research on Chemical Periodic Processes*,
SpringerBriefs in History of Science and Technology,
https://doi.org/10.1007/978-3-319-95108-9_1

Dictionaries give no reliable information about the author of the cited paper. He was called P. S. Munck af Rosenschöld and was related to a Swedish noble lineage. In the table of contents of the above journal, he is also named P. S. Munck af Nordenschöld.

The Swedish philologist Dr. Josef Eskhult explained to the present author the following: "The fact is that surname Munck af Nordenschöld is simply a mistake for the surname Munck of Rosenschöld, which is the name of a Swedish noble family. There is no historical person with the name P. S. Munck af Nordenschöld. In the table of contents, the author's name is mistakenly given as P. S. Munck af Nordenschöld. The paper or article was written by Peter Samuel Munck of Rosenschöld (1804–1860) as indicated on p. 216.

There is also a Swedish noble family by name Nordenskiöld, but Peter Samuel Munck af Rosenschöld seems not to have had any connection to the Nordenskiöld family".

Like the majority of experimentalists of the first half of nineteenth century, P. S. Munck had a wide scope of interests. For example, he was interested in electricity. "In the 69th volume of Peggendorf's Annalen, is a memories by Munck af Rosenschöld, in which induction is treated of. In his somewhat extended consideration of the subject, into which he naturally introduces much that is known, he starts with the correct view of induction, which is also defended Riess and Fechner" (Annual report…, p. 400).

R. Munck referred to Gustav Fechner in the above cited paper about the phosphorus light.

1.2 Gustav Fechner

Gustav Theodor Fechner (1801–1887) was Professor of philosophy and Professor of physics (1834) at University of Leipzig. His name is present in the books on the history of electricity. Fechner was one of the yearly specialists in electrochemistry.

Fechner wrote "Das Büchlein vom Leben nach dem Tode" ("The book of life after death")—1836. This book had been translated into English and published with Foreword written by W. James, who is among the classics of American pragmatism (1904).

Wilhelm Ostwald wrote in his autobiography: "Among the scientists we find Gustav Theodor Fechner to be one of the most original thinkers…

This unusual many-sidedness of Fechner was conditioned by the course of his development but also by his cast of mind. During his student days…Fechner came under the influence of natural philosophy. He has mentioned that his thinking was greatly influenced by the writings of Okean and Schelling…Later he studied the exact sciences. The need to earn his bread led him to take up translating French text-books… In these textbooks mathematical methods were successfully used. This activity soon led Fechner to perceive the unsatisfactory nature of natural philosophical fantasies. Fechner repeatedly mentioned that he had an inward sympathy for entire outlook of the natural philosophies" (Ostwald 2017, p. 468).

As it is said in Hedges and Myers' book (1926, p. 67), "Periodic electro-chemical phenomena it is convenient to consider in two sections:

a. Cases in which current is taken from the system.
b. Cases in which current is led into the system, this section may be further subdivided according to whether the phenomenon occurs at the anode or at the cathode.

The earliest case of periodicity of the first type was reported in 1828 by Fechner, who observed fluctuation in the potential of an iron electrode immersed in a weakly acid solution of silver nitrate and a corresponding periodic deposition and dissolution of silver. Similar observations were made a few years later by Schönbein, using iron in contact with copper".

R. Kremann provided more details (1913, p. 293): "Before his electrochemical studies Fechner studied the passivity of iron in nitric acid containing silver. Fechner stated that iron in silver solution is passive, and as some amount of nitrogen acid is added, iron is intensively attacked. Under some fixed amount of acid the sample is still active for a time, and then it becomes passive. Fechner observed that usually the processes did not continue long, the active state of iron changed for the passive state and back for its active state and so on four-six times".

Fechner showed that the chemical passivity is connected with electrochemical behavior.

"When he connected into a circuit together with a galvanometer two metals in an about 8%, acidly weakly reacting solution of silver nitrate and then observed the movements of the galvanometer, he found that there occurred repeated decreases and increases of the current, and even changes in its direction, in particular, if one of the metals was iron. In any case the phenomenon was too convoluted (verwickelt) to be sufficiently analyzed—due to the direct impact of the metals on the silver solution".

Wilhelm Ostwald discussed Fechner's 1828 paper at length in his book on the history of electrochemistry. However, in contrast to Hedges, Myers, and Kremann, he did not emphasize the phenomenon of chemical periodicity. He discussed this story in the context of the early development of electrochemistry, in the context of the collection of facts concerning electrochemical behavior. The point was an anomaly of electrochemical behavior of iron.

Ostwald gave an abriss of the prehistory: "In 1827 Dr. Gustav Wetzlar, a practicing physician in Henau, wrote: "If one pours a few drops of moderately concentrated solution of copper nitrate on the surface of a small bright bar of iron one will be astonished to see that the various drops behave in entirely different ways" (Weltzer's paper was published in Schweigger's *Journal f. Chemie and Physik.* vol. 49. 1827, p. 470)".

Ostwald wrote the following: "Direct measurement of the electric voltage or the direction of the electric current between the iron bar that had become negative and normal iron was not effected by Welzlar. This deficiency was soon made up by

Fechner (Fechner's paper in Schweigger's *Journal f. Chemie and Physik.* vol. 53.
1828, p. 141). Fechner fully confirmed Wetzlar's view on electrochemical state of
the changed iron. Fechner recounts that in a laboratory solution of silver nitrate the
was initially positive in relation to the silver. The iron soon became negative and,
finally, neutral. Fechner continues:

I repeated the above experiment with the portion of a laboratory silver solution
obtained elsewhere. The deflection given by the iron was likewise initially positive.
It soon changed to the opposite, negative. It remained continuously in the negative
position. I found that this silversalt solution showed a somewhat acidic reaction.
Therefore I added a few drops of a concentrated nitric acid to the neutral silver-salt
solution I had used before. In this case I obtained the same results. That is to say,
the deflection given by the iron was initially positive. It not only did not return to
zero but remained negative. When the deflection was negative I poured a very
considerable quantity of acid into the solution. I then repeated the experiment in the
way Wetzlar performed it. I immersed fresh iron and silver in the solution. The iron
immediately began dissolve and remained positive. The iron not only remained
bright but its negative deflection continued for some time. Then there was a sudden
dissolution of the iron and precipitation of silver. There was a simultaneous switch
from the negative deflection given by iron to a positive one. This was quite in
conformity with Wetzar's experiment and his views. The precipitated silver soon
disappeared. The iron again became bright and without action. At the moment this
happened the negative deflection by the iron reappeared. During the course of
repeated experiments I observed that the phenomenon rarely remained stationary.
Rather, the dissolution of the iron and its turning bright again along the dissolution
of the precipitated silver was repeated four or six times. This took place in quick
succession. Each time the magnetic needle was deflected in the opposite direction.
The process continued until the small bar of iron finally remained inactive on each
occasion" (Ostwald 1980, vol. 1, pp. 682–683).

In conclusion, Ostwald writes that Fechner discussed the possibilities of
explaining these phenomena. Fechner emphasized that his own attempts to find an
explanation were in vain.

It is interesting that in his 1832 three volume book "Repertorium der
Experimentalphysik: enthalten eine vollständige Zusammenstellung der neuen
Fortschritte dieser Wissenschaft" ("The main themes of experimental physics: the
containment of a complete set of the new results of this science") Fechner did not
mention his 1928 experimental result concerning "the dissolution of the iron and its
turning bright again".

1.3 Friedlieb Runge

About Friedlieb (1794–1867) there is considerable literature (Anft 1937; Bussemas
et al. 1994; Schwenk 2005; Ettre 2011; Niedobitek and Niedobitek 2010;
Schweitzer 1994; Senchenkova 1991, pp. 52–67; Stäudel and Wöhrmann 1998).

Runge's teacher was the famous chemist I. W. Döbereiner who formulated the law of triads. Runge had some achievements in analytical chemistry; usually he is mentioned as a predecessor of paper chromatography. He had achievements in organic chemistry, too. It is interesting that in English Wikipedia, he is characterized as a specialist in analytical chemistry, and in Russian Wikipedia, he is characterized as a representative of organic chemistry (chemistry of dyes).

Runge met some of geniuses of the nineteenth century. In particular, he demonstrated his results to Goethe.

J. R. Partington writes in his laconic stile: "F. F. Runge was at first a pharmacist, then associate professor in Breslau (after a long residence in Paris), then in the Prussian Marine service in Berlin and Oranienburg. He published several technological and other papers, also on the motion of electrolytically polarized mercury, and books. Runge rediscovered aniline in coal-tar oil and call it *kyanol* since it gave a blue color with bleaching powder, he found that it stained pine wood and elder pith yellow, and gave a greenish-black colour when oxidized by a cupric salt (aniline black). In the same research he discovered in coal-tar oil another base leukol (quinoline), an acidic substance which he called carbolic acid (kalboursäure, phenol), pyrrol (πυρρός, fiery-red), also rosolic acid, and three other bases. He did not analyse any of substances" (Partington, p. 183–184).

In 1850, Runge issued the book „*Farbenchemie. Musterbilder für Freunde des Schönen und zum Gebrauch für Zeichner, Maler, Verzierer und Zeugdrucker, dargestellt durch chemische Wechselwirkung*" (Color chemistry. Sample images for friends of beauty and for use by sketchers, painters, decorators, and printers, prepared by chemical interactions] (Berlin, (Germany): Self-published). A few libraries are keeping copies of the book. For example, the library of American Philosophical Society (Philadelphia) is keeping a copy of this book.

In 1855, Runge issued "Der Bildungstrieb der Stoffe: veranschaulicht in selbstständig gewachsenen Bildern (Fortsetzung der Musterbilder) (the formative tendency of substances illustrated by developed images)" (Oranienburg, self-published).

Bellow the title page of this book is represented. This is from the digital collection of Chemical Heritage Foundation (Philadelphia).[1]

---

[1]On February 1, 2018, the Chemical Heritage Foundation became the Science History Institute.

This copy is followed by the comment: "Early example of paper chromatography from German chemist Friedlieb Ferdinand Runge, who is considered to be the oregonator of the analytic technique. Runge investigated the colour reaction of various chemical substances such as ammonia and iron or copper oxide. The results include these 22 smaller chromatograms mounted on the title page surrounding the title and 60 larger cromotograms mounted on 31 leaves"

In this book, Runge was concerned with the enigma of his creative work. "I believe that the formation of the pictures is provided by the new unknown power. This power has nothing in common with electricity, magnetism, galvanism. It is exited from outside, but it leaves in substances, and it becomes active when substances become equal in their chemical opposition and this means binding through attraction and repulsion. I consider that this power is present in plants and animals, and I call it Bildungstrieb."

1.4 Quincke Georg

"Quincke was an experimentator of the highest range, for theories he had little affection" (Obituary, Nature, p. 280).

Quincke's former student Nobel Prize winner Ferdinand Braun wrote the following in connection with Quincke's seventieth birth day: "It is unnecessary here to discuss the value of the results of Quincke's work. Many of them have become an integral part of physics education, and have stimulated further development of physics. They have penetrated into all the spheres of physics, and many of his results, which were taken as doubtful at first, have been later confirmed, often by the researchers who did not accept them originally" (Braun 1988, p. VI).

Quincke wrote that there are considerable grounds for the view that periodicity in chemical reactions is associated with surfaces which are in metastable condition.

Mention is made by Quincke of the periodic spreading and accumulation of oil drops on a large water air surface such a pound.

"A bubble of air in water is trapped under a plane glass surface and kept under observation: on addition of alcohol the bubble contracts because of the alteration in surface tension. If the alcohol is added to the water in the vicinity of the bubble extremely slow from a fine capillary (so slowly that 1 c.c. is added in several hours) the bubble is observed to pulsate, contracting and expanding rhythmically in periods of 0, 1–10 s according to the speed of addition of alcohol. Precisely similar pulsations were observed in the case of an oil drop in water when sodium carbonate solution was allowed to flow very slowly from a capillary 0.1 mm in diameter in the neighborhood of the drop. The oil contained fatty acids with which the sodium carbonate could form a soap. The wave-length or time interval between the pulsation was observed to increase throughout an experiment" (Myer and Hedges, p. 85).

Here, the 1898–1902 papers have been cited.

In another paper published in 1902, Quincke was concerned with the periodic precipitation. "If two solutions of metallic salts in water produce chemical compounds which are insoluble in water, there originates (occurs) a precipitate. This precipitate takes considerable time to be formed and it can stay in the liquid medium for a while before its solidification.

Some amount of the compound is leaving in surrounding liquid as a solute. The precipitation occurs when supersaturated solution has been formed. A small amount of precipitate has been formed for the beginning. Then due to the contact mechanism all the compound has been precipitated.

Thus we have either immediate precipitation or periodical precipitation" (Quincke 1902, p. 642).

Quincke referred to Runge's writings. He did not look for a rational explanation of Runge's buildings and reproduced Runge's statement about the "life force" (ibid, p. 644).

References

Anft B (1937) Friedlieb Ferdinand Runge, sein Leben, sein Werk. Ebering, Berlin

Braun F (1988) Hermann Georg Quincke. Wied Ann 35:1–7

Bussemas HH, Harsch G, Ettre LS (1994) Friedlieb Ferdinand Runge (1794–1867): "self-grown pictures" as precursors of paper chromatography. Chromatographia 38(3–4):243–254. https://doi.org/10.1007/bf02290345

Ettre LS (2011) One hundred years of chromatography: a historical dialogue. Elsevier, Amsterdam

Frank-Kamenetskii DA (1939) Periodicheskie processy v kinetike okislitelnykh reaktsii (The periodical processes in the kinetics of oxidation reaction). Doklady AN SSSR 25:67–69

Gray P, Scott SK (1985) Isothermal oscillations. In: Field R, Burger M (eds) Oscillations and traveling waves in chemical systems. Wiley, New York

Hedges ES, Myers JE (1926) The problem of physico-chemical periodicity. Arnold & Co, London

Kremann R (1913) Die periodischen Erscheinungen in der Chemie. Sammlung. Chemischer Vorträge. Verlag von Ferdinand Enke 19:289–416

Niedobitek C, Niedobitek F (2010) Genie ohne Ruhm: Biographien von Walter Kausch, Franz Kuhn, Curt Schimmelbusch, Friedlieb Ferdinand Runge, Ernst Jeckeln, Friedrich Wegener. Jacobs Verlag, Lage

Ostwald W (1980) Electrochemistry. History and theory, vol 2. Smithsonian Institution by Amerind Publishing Co. Pvt of New Delhi, Washington, DC

Ostwald W (2017) The autobiography In: Scholz F, Jack R (eds) (trans: Jack R). Springer International Publisher

Quincke G (1902) Über die Klärung trüber Lösungen. Annalen der Physik. Bd.7:57–95

Schweitzer F (1994) Natur zwischen Ästhetik und Selbstorganization Theorie. Zum Naturbegriff der Gegenwart. Stuttgart-Bad Cannstatt. Bd.2:93–119

Schwenk EF (2005) Friedlieb Runge and his capillary designs. Bill His Chem 30(1):30–34

Scott SK (1991) Oscillations and waves in chemical kinetics, vol 90. Oxford Science Publications, Oxford, NY, Toronto

Senchenkova EM (1991) Rozgdenie idei i metoda adsorbtsionnoi hromotografii (The absorbational chromatography; its ideas and methods came into being). Nauka, Moscow, 228 pp

Stäudel L, Wöhrmann H (1998) Runge-Bilder aus der Expermentierkiste. Praxis Schule 5–10, 9. Heft 1, S.24–28

Chapter 2
Liesegang Rings and the Other Periodic Phenomena

Abstract Chapter 2 is about the discovery of Liesegang rings (Raphael Liesegang, 1896–97), it also concerns with the study of their modifications and search for their analogies in nature. The theories of Liesegang rings are also taken into consideration. It is shown that the discovery of Liesegang rings which are still attractive and enigmatic now helped to combine the varied research in chemical periodicity within a single research area. This chapter contains information on R. Liesegang's life and his creative work. It also referrers to scientists who contributed to the Liesegang rings studies: the German chemist Wilhelm Ostwald, his son Wolfgang Ostwald, the British chemist Hedges, the Russian university teacher F. Schemjakin and his student P. Mikhalev, the French chemist S. Veil, the Indian physicists C. Raman and K.S. Ramaiah, the Indian chemist N.R. Dhar. At the end of the chapter the mathematical modeling of Liesegang phenomenon is discussed from the historical point of view.

Keywords Photography · Catalysis · Colloid chemistry · Tube
Agate · Diffusion · Diffusion waves · Mathematical regularities
Quantum mechanics · Mineralogy · Curiosity · Commodity · Romanticism
Mathematical physics

2.1 Liesegang Rings

The systematic study of periodic structures can be traced back to Liesegang's 1896–97 publications (Liesegang 1896a, b, 1897). Liesegang reported that he covered a glass plate with layer of gelatin impregnated with potassium chromate and added a small drop of silver nitrate. As a result, silver chromate was precipitated in the form of a series of concentric rings, well developed and with regularly varying spacings.

© The Author(s) 2018
A. Pechenkin, *The History of Research on Chemical Periodic Processes*,
SpringerBriefs in History of Science and Technology,
https://doi.org/10.1007/978-3-319-95108-9_2

R. Liesegang (the photo is made in his seventieth birthday)

Hedges, who wrote a special book on Liesegang phenomenon (published in 1932), says that "the first serious study of a periodic structure was made by Liesegang in (1896a, b). If a drop of 10–20% silver nitrate is placed on a sheet of gelatin impregnated with 0, 4% of potassium dichromate silver dichromate is precipitated in the gelatinous medium. Under these conditions, however, the pre-cipitate is not continuous, but forms of series of concentric rings separated by clear spaces in the gel…

The periodic structures is by no means restricted to silver dichromate, no need to gelatin to be the medium, indeed one of the most important results emerging from a study of the literature over the period of thirty years is that the problem come is quite general one and there is every reason to believe that under suitable conditions periodic structures can be obtained through the relation of any two substances that form precipitate" (Hedges 1932, p. 13).

The review of Liesegang's first publications about his rings written by Wilhelm Ostwald and printed in the authoritative "Zeitschrift für Physikalisch Chemie" (1897) contributed to popularization of Liesegang's experiment. This review also contained the first theory of Liesegang rings. Wilhelm Ostwald postulated a metastability limit at which a supersaturated solution discontinuously precipitates out. He suggested that a precipitate is not formed immediately upon the concen-tration of the ions exceeding a solubility product, but a region of supersaturation occurs first. When the limit of stability of the supersaturation is reached, the pre-cipitate is formed, and a clear region is formed ahead of the diffusion front because the precipitate that is below the solubility limit diffuses onto the precipitate.

It is useful to reproduce some of Wilhelm Ostwald's definitions. They are present in his textbook. "If no sold is present, the concentration of the solution is not fixed. This is true whether the concentration is below or above the saturation value. Solutions which contain more of a solid than corresponds to saturation are called supersaturated" (Ostwald 1912, p. 327).

"If degree of supersaturation is increased (e.g. by cooling the solution of a substance, the solubility of which increases with temperature) a point is reached at which the solid separates spontaneously. We may then differentiate between a metastable region, following the stable region of supersaturation, and a liable region of greater supersaturation. It is very difficult to fix the boundary between these regions, as it has been shown to depend not only on the nature of substances but also on the presence of foreign solids (dusts) in a manner not yet known, none the less "metastable limits" have lately been experimentally proved to exist" (ibid, p. 328).

In 1907, Liesegang conducted an experiment which seemed to refute Wilhelm Ostwald's theory (Liesegang 1907). "The author brings forward some results that are not in agreement with the supersaturation theory" (Chemical Abstracts 1908). However in 1911, he noticed that his experiment can be explained from the point of view of Ostwald's theory (Liesegang 1911).

By considering the history of Liesegang rings, one should have in mind the following: "Liesegang's structures were called rings, because they were first observed in that form, i.e. as concentric deposits in a plane. Later they were more often grown in test tubes, and were therefore disks, but the 'ring' name was by then firmly implanted in the literature, and has prevailed to this day" (Henisch 1988, p. 2).

A lot of experiments with Liesegang structures followed Liesegang's 1896 publication. Hedges–Myer's 1926 book and Hedges (1931, 1932) books provided the review of results obtained before 1931 (it should be noted, however, that these books belong to the class of rare books). Henisch' 1988 book contains more extensive list of experiments (the formation of rings by means of organic substances, the formation of rings in the presence of electric fields, etc.). "Rings were considered interesting partly because their origin was obscure and partly because they reminiscent a certain structure found in nature, e.g. the striation of agate" (Heinz. p. 2). True, "it is necessary to admit that the study of this beautiful phenomenon... has not yet found its practical niche" (Heinz. p. 116).

Many discussions concerning the natural analogues of Liesegang rings followed Liesegang's discovery. First of all, the rings found in agates and malachite were taken under consideration. The methodology of Liesegang ring became to penetrate into some spheres of biology. However, it is doubtful to apply the methodology of artificial Liesegang rings directly to the natural structures. As Hedges wrote, the "periodic ring and layer formations found in nature offer only the most limited opportunities for research into their origin. Indeed many are due to quite different mechanism, e.g. changes of overall environment, ever though they bear the superficial appearance of Liesegang ring" (Hedges 1932, p. 87). At the same time, Hedges emphasized that the physico-chemical processes which lead to Liesegang phenomena can be used to construct useful models in the course of research of the natural processes which lead to the analogues of Liesegang structures.

Liesegang himself actively participated in the attempts to understand the mechanism of the formation of the natural structures similar to Liesegang rings: He tried to apply his 1896 methods but he was accurate enough to avoid a direct extrapolation.

In 1913, the German botanist E. Küster published his results concerning rhythmic structures in the tissues of plants (Kuster 1913). Lizegang (1913) was in hurry to discuss these results in terms of his methods, which can be traced back to his 1896–97 papers. But he turned to the concept of supersaturated solution used by Wilhelm Ostwald in his theoretical explanation of Liesegang rings, and to the concept of nucleus (Kern) to explain the precipitation (instead of Ostwald's supersaturation limit).

In 1913, Liesegang also published a book explaining geological periodical structures (agates, malachite, etc.). He referred to the processes of diffusion of some salt solutions which reacted with each other to form a precipitate. Agates probably arose within cavities in solid masses with which they were not genetically connected. Hot streamers brought the solutions which filled the cavities. In the course of time, these solutions were transformed into gels.

In 1915, Liesegang published a special book on agates. In this book, he described his experiments which allowed him to produce the artificial agates.

Liesegang's discovery induced not only a series of experiments. A number of theories which explained Liesegang rings arose.

We mentioned Wilhelm Ostwald's theory. It was criticized by his son Wolfgang who proposed another theory.

Before Wolfgang Ostwald, an alternative theory was proposed by the British scientist Samuel C. Bradford. This theory attributes the spaces between the rings (the bands) to the adsorption of the solute in the gel onto colloidal particles of the precipitate which is formed. According to Bradford, the diffusion of silver nitrate initially results to the sol of the substance which should form the precipitate. In the course of time, the precipitate appears (the coagulation of sol takes place) and it adsorbs substance which is situated near the precipitate. As a result, a clear space is formed near the precipitate portion.

The concept "diffusion" is present in the above-mentioned theories. The theory developed by Wolfgang Ostwald (1922) put this concept at the forefront.

Wolfgang Ostwald postulated the existence of three diffusion waves in the system which results in the Liesegang rings: (1) the diffusion wave of the external electrolyte, (2) the diffusion wave of the inner electrolyte, and (3) the diffusion wave of the soluble product of reaction. He also postulated that the precipitation is run by the law of mass action. Let us take the following reaction:

$$2NH_4OH + MgCl_2 = Mg(OH)_2 \downarrow + 2NH_4Cl$$

The law of mass action leads to the formula:

$$[Mg(OH)_2] = k \frac{[MgCl_2][NH_4OH]^2}{[NH_4OH]^2}$$

The magnesium chloride and ammonia diffuse toward each other and react to give insoluble magnesium hydroxide and soluble ammonium chloride. The ammonium chloride is produced at high concentration and has a high ability to diffuse (it diffuses ahead of the ammonia). What follows the formation of a precipitate? The area of the gel is formed, the area containing high concentration of ammonia chloride and relatively low concentration of magnesium chloride. When ammonia reaches this region by diffusion, we have the situation in which in accordance with the law of mass action magnesium hydroxide is not precipitated. However, a little further, as a result of the delusion of ammonium chloride and the still high concentration of magnesium chloride, the relation of concentrations leads to the situation where magnesium hydroxide is precipitated. The reaction in this place produces more ammonium chloride which diffuses ahead and again prevents the precipitation in the next area. The situation is again reproduced.

As a development of Wolfgang Ostwald's theory, the Soviet chemists (in the contemporary terminology, Russian chemists) Fedor Schemjakin (1905–1981) and Pavel Mikhalev (1908, the present author was not able to find the year of his death),[1] who both worked for Lomonosov Moscow State University, proposed their emission-wave theory (1934–35). Some of the journal papers presenting this theory had been written by them in coauthorship with W. K. Nikiforov, who was a researcher at Belorussia State University. All the important articles had been published in Kolloid-Z, edited by Wolfgang Ostwald.

"The specificity of periodic reactions implies the idea that they cannot be described by the ordinary waves" (Schemjakin and Mikhalev 1938, 118). Nevertheless, if light can display wave and particle properties, then the matter can also have a wavelike nature. The Schemjakin–Mikhalev theory tends to be general. They treated the wave of diffusion as matter waves in the L. de Broglie's fashion. As is known, L. de Broglie proposed that if there exists a particle which has energy E and momentum whose module equals p, then the wave is associated with this particle. In other words, the formula $p = \frac{h}{\lambda}$ is valid (the wavelength is λ, and h is the Planck constant).

As Schemjakin and Mikhalev highlighted, that had been expressed by the following statement of the French biologist and philosopher S. Leduc (1853–1939): "It was said that the phenomena of interference and diffraction could not be explained by the theory of emission, while the undulatory theory gave a simple explanation. The scientific mind was unable to conceive the idea of emission and periodicity as taking part in the same phenomenon. The savants and thinkers who have meditated on this question have always considered the theory of emission and that of periodicity as incompatible. Nevertheless, we are here in presence of a phenomenon in which emission and periodicity exist simultaneously" (Leduc 1911, p. 91).

Here, Schemjakin and Mikhalev cited Stéphane Leduc's *The mechanism of life* which was first published in French and have then been translated into English (Leduc 1911) and into German.[2]

---

[1]The Archives of the Frumkin Institute of Physical Chemistry and Electrochemistry, where Mikhalev worked after postgraduate studies at MSU, do not contain any information on Mikhalev. His last paper was published in 1939 (Mikhalev et al. 1939).

[2]It should be noted that its English translation appeared in Internet in 2010: http://www.gutenberg.org/.

Schemjakin and Mikhalev were very far from the construction of a mathematical deductive theory. They discussed their problems by appealing to analogies and even metaphors (although they themselves did not emphasize this feature of their approach). "By applying the de Broglie equation to the periodic reactions we take the distances between the middle lines of the layers as the wavelength, the velocity of the diffusion field as the particle velocity (they meant the velocity which enters de Broglie's formulae—A. P.), the mass of moving particles as mass" (Schemjakin and Mikhalev 1938, p. 101).

As one of their achievements, Schemjakin and Mikhalev consider the description of the "starry rosettes" (the patterns resembling a star in shape) which can be observed by studying minerals. They can also be artificially produced (see: Schemjakin and Mikhalev 1938, p. 15). In this connection they turned, once again, to Stéphane Leduc who probably was the first to reproduce and describe this phenomenon. Schemjakin and Mikhalev themselves produced the starry rosette for the silver dichromate stratification in gelatin.

To confirm their approach, Schemjakin and Mikhalev refer to the empirical description of five reactions in their earlier published paper. What follows is the summary of their (together with Nikiforov) 1934 paper: "For five periodic reactions in gels it has been shown that the product of two values—the distance between the layers and the propagation speed of the diffusion field—is a constant value $\lambda v = \text{const} \left[\frac{cm^2}{sec} \right]$. This constant, if it is calculated over a long period of time, can be taken as a suitable and subtle quantitative characteristic of a periodic reaction" (Mikhalev et al. 1934, p. 167).

In particular, they considered the following reaction in gelatin:

$$2Ag\,NO_3 + K_2Cr_2O_7 = Ag_2Cr_2O_7 + 2KNO_3.$$

If gelatin contains 0005 M dichromate, the "constant of periodicity" oscillates between 3,3 and 3,8 $\left(\text{const} = \lambda v 10^6 \right)$. Under some other conditions, they came to less dispersed values of the constant.

However, how to apply de Broglie's wave–particle scheme to the waves of matter? Here, Schemjakin and Mikhalev turned to the statistical (ensemble) interpretation of quantum mechanics put forward by the Soviet physicist Nikolsky (1934, 1936, 1937). The story of ensemble interpretations of quantum mechanics is long and controversial (see: Pechenkin 2012). However, Schemjakin and Mikhalev emphasized only one point: Liesegang rings are a macroscopic quantum effect which can be explained if quantum mechanics is treated as a theory of ensembles of particles rather than particles (Schemjakin and Mikhalev 1938, 124). Instead of the wave–particle duality, they formulated another principle: the wave–ensemble duality.

A lively interest in Liesegang rings can be followed along K. H. Stern's bibliography (1955). The second edition of it was published in 1967.

2.2 Liesegang's Biography

There is a short record in the English translation of Wilhelm Ostwald's autobiography (a comment of the editors): "Raphael Liesegang (1869–1947). German chemist and writer. He did not have a university degree. After a course in analytical chemistry with Carl Remigus Fresenius (Wiesbaden) and studies in Freiburg, he worked in the family photographic factory. He discovered periodic precipitation rings in gels, advanced the photographic development process and contributed to colloid science".

There is an outline of Liesegang's biography in Henisch's book which has been cited above, there is a short comment in Niederson–Kuhnert's book (1987). There is a piece in "Neue Deutsche biographie" (Vol. 14, 1985, p. 535, there is online version). There are several notes concerning Liesegang's life and his creative work in the Internet (see: Beneke K. Liesegang named in literature; Kolloidwissenschaftler Raphael Liesegang).

On the occasion of Liesegang's 70th birthday, his friends and colleagues issued a Liesegang Festschrift under the editorship of Wilhelm Ostwald (see: Heinz, p. 118). This volume likewise reflects the great variety of his interests. Wilhelm Ostwald wrote the introduction and referred to Liesegang's 'incredible universality'. He noticed that there were the difficulties of providing an adequate overview of such a man's work. He had, indeed, been stunned by that task on a similar occasion ten years earlier and then asked Liesegang for a self-assessment, which Raphael Eduard Liesegang duly provided "to prevent other people from spreading false notions about me" (Liesegang 1929, p. 226). This autobiography, says Wilhelm Ostwald, is itself a typical piece of Liesegang.[3]

"'I created a commotion from my first schooldays onward', he begins and goes on to explain that he was always a stubborn and unresponsive pupil, as long as he found himself in situations in which learning was an imposed obligation". "I am glad I never had to teach, it allowed me to remain a student through all my life", Lisegang wrote.

According to the Wilhelm Ostwald's famous classification, Raphael Eduard Liesegang was a romantic. According to P. Duhem's classification, his mind was the broad-shallow English mind. It was never easy to classify him into any professional niche. "In fact, Raphael Liesegang was as much a physicist as he was a chemist … He did many things in his lifetime that would make a modern scientist recoil in awe … Thus, for instance, he was active as a bacteriologist, contributed to the chromosome theory and to the beginnings of paper chromatography, not forgetting the properties of aerosols and gelatins, the origin of silicosis, the mechanism of the photographic process in black-and-white and color (1889). When he was 29 years old, he published the book on chemical reaction in gels (1898), a topic to which he was drawn by his even earlier interest in photography (see also Liesegang 1896a, b). This interest was in the blood, so to speak, since J. P. E. Liesegang, his

---

[3]This autobiography is contained in (Beneke).

father, and F. W. E. Liesegang, his grandfather had both been pioneers of photography and the photographic industry" (Henisch 1988, p. 116–117).

It would be injustice to take forgiving tone by describing Liesegang's biography. Liesegang was the author of some fundamental books. His books on geological diffusion and on agate were mentioned above. In 1915, he published a book on colloid chemistry. True, in the preface he announced that his book is not a textbook. He writes that he offers the reader a must in the process of fermentation, and not a ready-made wine.

Liesegang was the editor of a collective Grossbuch on *Kolloidchemie Technologie* (1927). Thirty authors participated in composing this book. Liesegang himself wrote not only Introduction, but three fundamental texts: Glass (p. 634–674), Photographie (p. 979–1008); Conclusion (p. 1009–1029). Liesegang regularly published his reviews in *Kolloid-Zeitschrift* edited by Wilhelm Ostwald.

2.3   Wilhelm Ostwald

The chemical periodicity was one of hundred problems with which Wilhelm Ostwald was concerned. Among 45 books written by Ostwald, there is no book about periodical processes. Nevertheless, he mentioned this problem in his three volume autobiography "Lebenslinien." He wrote about chemical periodicity in chapter entitled "Overload, breakdown, recovery. To be more specific, he wrote about chemical periodicity in section entitled "Farewell to the bench. This means that Ostwald had been very tired and he made his mind to say "good bye" to his former hard and in part routine work as a laboratory researcher and to turn to basically organizational and theoretical problems. The two following chapters are entitled "Society of electrochemistry" and "Institute of catalysis".

"I got a little solace from a different experimental work I carried out in 1899.

W. Hittorf had noticed some curious properties of a sample of metallic chromium which he'd got from H. Goldschmidt. He gave me a small sample so that I could look at it myself and I found that when it was dissolved in hydrochloric acid the release of hydrogen soon stopped and then started again vigorously without there being a change in any other properties. When I looked more closely it became clear that this behavior was periodic and I showed using a stop watch that the periodicity was constant. This spontaneous periodicity fascinated me because I'd already in another context—the periodic precipitations in gelatin discovered by R. Liesegang and known as "Liesegang's rings"—come across the question of how periodic behavior can be generated in a situation where all the experimental conditions are constant. At that time I had a reasonable explanation but it was tailored to this particular problem and did not provide a general explanation of such events and such an explanation was now clearly required".

Ostwald mentioned Johan Wilhelm Hittorf (1824–1914) and Hans Goldschmidt (1861–1923). Hittorf is known by his work on the mobility of ions.

Ostwald had in mind his explanation of Liesegang rings, which he provided in his 1897 review of Liesegang's 1896 paper (see above). This explanation was reproduced in Ostwald's great *Lehrbuch der Algemeinen Chemie* (Zweiten Bandes zweiter Teil. Verwandtschaftlehre).

Here, Ostwald presented his theory of Liesegang rings in connection with the problem of experimental estimation of the supersaturation limit. He wrote that the chemist who attempted the problem of supersaturation limit by cooling the solution was not successful.

The key sentence was "The sediment does not appear immediately, it appears as metastability limit has been reached" (Ostwald. Lehrbuch…, p. 778).

In his autobiography, Ostwald came back to his experiments with the periodic evolution of hydrogen. "The first thing that had to be done was to decide on a procedure with which the necessary measurements could be carried out precisely and in the minimum possible time. I couldn't bring myself to burden an assistant with making the boring measurements and recording the results. On thinking it over I wondered if the phenomenon might not record itself along the lines of the principle introduced into physiology by C. Ludwig. I was conversant with the equipment he'd used from my visits to the physiological institute. I soon found the solution which was to use an elastic capsule whose movements could be recorded by a lever holding a pen that drew a line on a moving strip of paper. The necessary pressure difference was achieved by using a capillary to slow the outflow of the hydrogen. In a short time an apparatus was designed and built which with little effort was able to record six parallel determinations run for as long as one wanted and deliver the results in graphic form" (Ostwald 2017, p. 265).

Carl Friedrich Wilhelm Ludwig (1816–1895) was a German physician and physiologist. "Carl Ludwig is the greatest physiologist of his time", Wilhelm Ostwald repeatedly writes in his autobiography (p. 147, 151, 195, 196).

"Once again I felt happy at having found a good solution to the technical problem. As far as the science went, however, though we found a number of regularities they did not lead to a general explanation. Part of the problem was that this property of chromium was seen only in the first sample. All later samples dissolved without periodic release of hydrogen. I asked for and got from H. Goldschmidt numerous further samples of chromium but none of these showed the phenomenon. So, once the first material was used up we had to abandon the project" (ibidem).

Ostwald published his results in 1899–1900. «A better known example is Ostwald's research on the periodic evolution of hydrogen accompanying the dissolution of chromium in acids.

The rate of the reaction was followed by employing a "chemograph" which registered automatically and continuously, the pressure of hydrogen evolved. The first experiments with chromium showed that the rate of dissolution varied periodically and was accompanied by alternate periods of rapid and slow evolution of hydrogen, which Ostwald supposed to be due to alternating periods of activity and passivity of the metal…

A most important observation made in this work is that when the particular batch of chromium was used up, no periodic results could be obtained from other specimens... Fortunately Goldschmidt was able to provide a further small supply of the original batch of metal and this enabled Ostwald to show that periodic E. M. F. was given by the cell Cr (periodic)/HCl/Pt» (Hedges and Myers 1926, p. 39–40).

Although the chemical periodicity was one of the many topics which Wilhelm Ostwald followed through his life, his contribution to this subject was outstanding. Ostwald formulated the concept of autocatalysis, which turned out to be essential when the mechanism of the periodical processes has been taken under analysis. Alfred Lotka who was the founder of studies in mathematical foundations of kinetics of the chemical periodical reactions spent several months at Leipzig University in 1901–1902. As Lotka's biographer writes, Ostwald's lectures and his physical chemistry strongly influenced Lotka (see: Lambert M. Surhole. Alfred Lotka, http://fr.academic.ru/dic.nsf/frwiki/80720; Kingsland 1985). As a whole, Wilhelm Ostwald had an impact on the development of chemistry in the USA (Servos 1990).

2.4 Suzanne Veil

Suzanne Veil's biography can be found in "The biographical dictionary of women in science". Vol. 2, p. 1325 (Marilyn Oglivie, Joy Harvey (eds). Routledge. NY, London 2000).

She was born in 1886, the year of her death is unknown. She received Doctor Science degree in 1920; in 1924, she became a Professor at University of Paris.

Veil published numerous articles on various aspects of physical chemistry. She distilled some of her results into her 1934 publication on periodicities in chemistry (Veil 1934, part 1 and part 2). In addition to drawing inferences from a wide range of chemical phenomena, she included references to biology and geology in this work.

S. Veil's research embraced many different periodic phenomena. However, her interest often concentrated on Liesegang rings. She took the kinetics of the ring formation under consideration by applying a variety of physico-chemical methods which were unusual for the field.

In 1930, she published the article entitled "A photomicroghic study of Liesegang rings" (Comp. Rend. 1930. V. 191, p. 611–612). Here is the abstract of this paper. "A gelatin soln impregnated with $K_2Cr_2O_7$ was spread thin on a glass plate and a single drop of $AgNO_3$ soln was placed on this plate. The Liesegang rings that developed were examined photomicrographically. The principal rings were found to be so spaced that the square roots of the distance of between successive rings varied in a simple arithmetic series. This quantity when plotted against the numerical order fell closely on straight line".

In 1931, Veil published in coauthorship the paper entitled "Microscopic and cinephotographic study of Liesegang rings" [Veil and L. Bull. Compt. Rend. 192, 282–4 (1931)].

The rings formed by the action of a drop of $AgNO_3$ on dichromate impregnated gelatin formed concentrically with pointed protuberances interrupting their continuity. As successive rings were formed, the protuberances receded but the rings were left incomplete at these points. The velocity of formation of the rings decreases and the intercepting protuberances disappear as the contact line between soln recedes toward the center.

A summary of Veil's 1932 paper "The individual diffusion of Liesegang reactants in gelatin" (Compt rend. 194, 1155–57. 1932) says the following: "A drop of $K_2Cr_2O_7$ solution on gelatin diffused so that no limiting border could be observed while a drop of $AgNo_3$ solution diffused with a well-defined limit. Two drops of these reagents when allowed to diffuse together formed symmetrical stratification but under the influence of electric potential symmetry was disturbed or altered".

S. Veil was one of the scientists who started to investigate mathematical regularities in the structure of Liesegang rings. True, these were empirical regularities. S. Veil followed Schleussner's 1922–1924 papers where the law concerning the distance between successive bands was proposed. It was known that the distance increases as the diffusion proceeds. H. Schleussner showed that the relation of the distances of two any successive bands equals a constant (the "distance" means a distance from the band to the center).

$$\frac{y_n}{y_{n-1}} = q$$

y_n, y_{n-1} are distances of two successive rings from the centrum. This means that the distances between successive bands grow as a geometric series (Schleussner 1922, 1924).

S. Veil's empirical formula connected the number of a band with the square root of the distance between two successive bands (Veil 1934, Part 2).

S. Veil extended her research on spiral-like precipitates (see: Veil 1934), she took the "starry rosettes" under consideration. As starry formations disappear quickly sometimes, she used cinema methods to study them (Comp. rend. Vol. 199, p. 282).

2.5 Liesegang Rings in Russia

Liesegang rings gained some popularity in the Soviet Union. Russian Orthodox Church priest and philosopher Pavel Florensky several times mentioned this phenomenon in his letters to his son who wanted to become a geologist (1935–36).

In 1928, Florensky who lived in Zagorsk (nearby Moscow) was exiled to Nizhny Novgorod. In 1933, he was arrested according to the Article 58 (agitation against the Soviet power) and sent to Baikal–Amur Mainline camp. In 1934, he was moved to Solovki. He worked there for a scientific division and conducted research into producing iodine and agar out of local seaweed. In 1937, he was transferred to

Leningrad (Sankt Petersburg now) and he was sentenced to death. He was buried in the common grave together with thousands of political prisoners.

Florensky could read scientific literature when he worked at Solovki camp. In his letters, he attempted the applied significance of Liesegang rings. He proposed that ages for a rock can be dated with the help of Liesegang phenomenon.

Nesterov. Philosophers. Pavel Florenskii (on the left) and A.Bulgakov

"Liesegang rings" is the title of the book written by the famous politician, the leader of the political party Spravedlivaya Russia (it can be translated as "Russia for justice") Sergei Mironov (2012). Mironov is a geologist by training and worked as a geologist for a while. However, this book is not about geology, mineralogy, and related topic. This book is about politics and this is the life story of a politician.

Let us turn to the people who really worked on Liesegang rings in laboratories (see also: Pechenkin 2017). At the second half of the 1920s, F. M. Schemjakin, who was mentioned in sec. 1, was a teacher at the Moscow State University and worked for the Research Institute of Chemistry which belonged to the Moscow State University (in 1933, he became docent), started his experiments with Liesegang rings. Schemjakin cooperated with Mikhail Semenovich Dunin (1901–1993), who, like Schemjakin, was a student of the Moscow State University, but, in contrast to Schemjakin, had not graduated from the University; Dunin left the University when he was a four-year student. When Dunin cooperated with Schemjakin, he was the chief of the agricultural laboratory organized at the newspaper named "The poor."

It is remarkable that Schemjakin and Dunin cooperated not only in the field of Liesegang rings, they studied the agricultural problems together (see Dunin and Schemjakin 1928a, b; Schemjakin 1931). However in the 1930s, they parted their scientific ways (see a booklet: *Mikhail Semenovich Dunin*, 1986, see papers: 'On Birthday of Fedor Mikhailovich Schemjakin' 1965, 1975).

Schemjakin Fedor Mikhailovich (the 1960s)

Pavel Fedorovich Mikhalev helped Schemjakin, and he was Schemjakin's graduate student. He graduated from the Moscow State University in 1931. In 1935, he received Ph.D. from the Moscow State University (the title of his thesis: "On the periodical reactions"). Since 1934 he worked as a researcher for the Colloid-Electrochemical Institute belonging to the USSR Academy of Sciences (later this Institute was renamed as Institute of Electrochemistry, now Frumkin Institute of Physical Chemistry and Electochemistry).

Schemjakin, Mikhalev, and Dunin mainly published their papers in *Kolloid Zeitschrift* (Koll-Z) founded by Wolfgang Ostwald. They also published their papers in the leading Soviet journals—*Journal of General Chemistry* (ZhOCh), Soviet Compte Rendu (*Doklady Akademii Nauk*), *Journal of Physical Chemistry*. In Koll-Z, the articles which reacted to their articles also appeared.

F. M. Schemjakin and M. S. Dunin wrote about the secondary system of Liesegang rings in a tube (Dunin and Schemjakin 1926). They described the classic interaction of silver nitrate and potassium dichromate, the interaction which results in the silver dichromate precipitate.

"In 1925 May, we conducted experiments concerning the formation of periodic bands of precipitate of silver dichromate. They were carried out in glass tubes of 50 cm length and 6–7 mm diameter. Initially, the process of the formation of precipitate had its usual appearance. First, about the upper end of the glass tube the continuum of layers of the precipitate was formed in the dilute gel under gelatin. Then a discrete sequence of layers of the precipitate appeared. The following stage was the formation of the silver dichromate globules. For the beginning the globules were so small that one needed a microscope to see them. Then the globules increased in their size. The globules are evenly distributed. With all these the process came to an end.

After infinitely long time (several weeks) a new process of the band formation started and the above picture arose again. A new system of rings started beyond the first system and passed through the same stages of formation. However, the new rings were smaller".

Boris Aristarkhovich Dogadkin (1898–1975), who criticized Dunin and Schemjakin's article, also graduated from Moscow State University. However, since 1929 he worked for Research Institute of Rubber Industry.

Dogadkin explained the formation of the secondary system by impurities (Dogadkin 1928b). He wrote that the current theories of the Liesegang phenomenon were not able to explain the secondary system of rings by proceeding from the internal reasons of the phenomenon. The secondary periodicity could only be explained by the action of external factors that produce synchronous changes in the process of precipitation.

Dogadkin referred to the anomalous rings which had been explained by the intermittent illumination. As a matter of fact, the various external factors affecting of solubility of the product during the course of diffusion occurred to be reasons of the additional anomalous rings.

Dogadkin had his own line in the discussion of the Liesegang phenomena [see his articles in Koll-Z, his review (Dogadkin 1926) and his small book (Dogadkin 1928a)].

However, Dunin and Schemjakin published the second paper where they insisted that they discovered the formation of the secondary system of Liesegang rings (Dunin and Schemjakin 1929). Their approach has been supported by E. Hedges who writes in the above-cited book (Hedges 1932, p. 31):

"Further experiments have been conducted by Dunin and Schemjakin... The experiments were carried out by allowing the reagent to diffuse into tubes 150 cm. in length over the period of a half years, the tubes being kept in dark meanwhile. The complex periodicity was observed with silver chromate, silver phosphate, and lead iodide. The authors suggest that the precipitation is controlled by three critical concentrations, each of which gives rise to its own periodicity, and they further suggest an analogy between the existence of "dead zones" exhibited in Liesegang ring formations and the similar "dead zones" which have been observed in the reception of radio waves and the sound of gunfire".

E. Hedges writes that the idea that one periodicity may be superimposed on another recalls some observations of his and Myers on the periodic dissolution of metals in aqueous solutions. He means the experiments published in 1924–25.

Schemjakin and Mikhalev published a number of papers concerning the morphology of Liesegang rings. As one of their achievements, Schemjakin and Mikhalev consider the description of the "starry rosettes" (the patterns resembling a star in shape) which can be observed by studying minerals (see the first section of this chapter).

2.6 Liesegang Rings in India

The present author has not any date on the reaction of Indian philosophers to Liesegang's discovery (let us recall that the Russian philosopher Pavel Florensky wrote about Liesegang rings in his letters to his son). However, as it follows from references in Myer and Hedges book and in Hedges books it is clear that Indian chemists and physicists contributed to the experimental and theoretical studies concerning Liesegang ring. An Indian specialist in physical chemistry and biochemistry Nil Ratan Dhar (1892–1986) wrote in coauthorship a number of papers on kinetic of the formation of Liesegang rings (Dhar and Chatterji 1925a, b; 1928). Hedges writes that they discriminated between two kinds of ring formation: "in one class layer of precipitate is followed by clear zone, in the other class a coagulated sol is followed by zon of peptized sol. They believe that periodic structures are formed only when the gel has a medium of peptizing influence on the precipitate" (Hedges 1931, p. 233).

True, the main contribution of Dhar to physical chemistry was lying in the field of photochemical nitrogen fixation. His main results were published at the beginning of the 1930s and in the middle 30 s.[4] However, in the course of his trip to Germany in 1926 he visited Wolfgang Ostwald's laboratory in Leipzig and had long discussions with Wolfgang Ostwald. He also had a meeting with R. Liesegang at his home in Leipzig (insandia.res.in/BM/BM14_8901.pdf). Judging on Dhar's publications, it is probable that he discussed the theory of Liesegang rings with Wolfgang Ostwald.

---

[4]Nil Ratan Dhar is considered to be one of the founders of physical chemistry in India.

The Nobel Prize Winner C. V. Raman also contributed to the study of Lisegang rings. Here, two articles published in 1939 (Raman–Ramaiah article and Ramaiah article) should be firstly mentioned. C. V. Raman and K. Subra Ramaiah published an article dedicated to the experimental proof of the suggestion regarding the wavelike character of Liesegang's precipitates. They wrote in this article that "when it is completely formed, a Liesegang pattern is a static structure and therefore scarcely to be regarded as a wavelike phenomenon in the usual sense, as the latter involves a movement or periodicity. It may be permissible, however, to describe a periodic precipitate as a wavelike phenomenon, meaning thereby, that it presents some analogies in its spatial distribution to the configuration at a particular instant of a periodic train of waves" (Raman and Ramaiah 1939, p. 467).

It is interesting that C. V. Raman and K. S. Ramaiah referred to the Schemjakin–Mikhalev–Nikiforov theory as it was presented in their journal papers. Raman and Ramaiah considered that Mikhalev–Schemjakin–Nikiforov ideas support their proposition: They applied the idea of diffusion waves to interpret the important feature of the observed pattern, namely the widening of space between of successive rings which corresponds to the slowing down of the diffusion occurring as we proceed outwards from the centre of the pattern. C. V. Raman and K. Subra

Ramaiah also supported the idea of the physical connection between the formation of the periodic precipitate and the de Broglie matter waves associated with the diffusion ions.

2.7 Liesegang Rings from the Point of View of Mathematical Physics

Liesegang rings attracted attention of the Soviet great specialist in mathematical physics Academician Yaakov Borisovich Zeldovich (1914–1987). Zeldovich had been awarded three times the gold medal of the Hero of Socialist labor for his contribution to the development of nuclear military technology; the head of Soviet nuclear team Igor Kurchatov also had won three golden medals; only Andrey Sakharov had won four such medals.

Zeldovich presented his conception of Liesegang rings in his paper written in coauthorship and published in *Doklady Akademii Nauk* in 1961. This article had been cited in English language papers on the theory of Liesegang rings. However, this article followed a couple of Zeldovich's papers published in the "Zhurnal Fizicheskoi Khimii" in 1949 (these papers have also been written in coauthorship, but with the other coauthors). These articles had no considerable resonance in scientific literature. However, in these articles mathematics which would be used in Zeldovich and coauthors' 1961 article had been under preparation.

As was noted above, Wilhelm Ostwald's 1897 theory became the first theory of Liesegang rings. Wilhelm Ostwald's theory of the diffusion waves and Schemjakin–Mikhalev's emission-wave theory followed it. Zeldovich and his coauthors came back to the first theory.

Zeldovich and his coauthors ideas were developed by several physicists. As was noted in Smith's article (Smith 1984), Zeldovich and coauthors treated Wilhelm Ostwald's theory of supersaturation in the terms of mathematical physics. "Ostwald's ideas were not given quantitative form for half of a century. Mathematical formulation of the Ostwald mechanism was given by Prager (German-American professor in applied mathematics, 1903–1980) in 1956 and by Zeldovich et al. (1961) for the special case where the initial concentration ratio is very large. They predict that the spatial spacing $\Lambda x_m = x_{m+1} - x_m$ and the time spacing $t_n = t_{n+1} - t_n$ satisfy asymptotic relations

$\frac{x_{n+1}}{x_n} \approx 1 + \xi_*$, $\frac{t_{n+1}}{t_n} \approx 1 + \tau_*$, where ξ_*, τ_* are constants" (Smith 1984, p. 3102).

Smith states that Prager and Zeldovich's results "correctly predict many features of the process of ring formation in some systems" (Smith 1984, p. 3102). Smith refers to Prager 1956 paper and to the English version of Zeldovich, Barenblatt, Salganik 1961 paper.

By starting their 1949 paper, Zeldovich and Todes (Oskar Todes is a specialist in mathematical physics, Zeldovich's friend) wrote "Now it can be taken as firmly confirmed, that the periodical structure of deposits results from the ability of solutions to take a supersaturated state. We shall analyze the reaction of two solutes (a and b), the reaction which results in the formation of non solvable precipitate.

Originally these substances are spatially separated, for example a is located on the left from the origin, and b is located on the right from the origin. The formation of precipitate depends of the mutual diffusion of a and b. If the supersaturation has not been attainable, a and b would not be able to exist in the one point simultaneously, sedimentation would not occur. Under this circumstance, at any given time sedimentation should take place on the boundary which separates two areas, the area of substance a and the area of substance b. The boundary where sedimentation takes place moves in space. The direction of its motion depends on which substance diffuses faster" (1949, p. 180).

It is probable the Liesegang rings problem attracted Zeldovich and his colleagues when they read Schemjakin–Mikhalev's writings. Zeldovich and Todes wrote that in Schemjakin and his collaborators established the important regularities: The distance of the precipitate layer from the point of the initial solvent interaction is proportional to the square root of the time, and (2) the distance between successive bands grows as a geometric series. As a matter of fact, these regularities have been found by the other chemists. However, Schemjakin and Mikhalev emphasized their importance for the theory of Liesegang rings.

Zeldovich   and   his   coauthors   severe   critisised   the   Schemjakin-Mikhalev-Nikiforov emission-wave theory. Zeldovich and Todes write decisively: "In his theoretical paragraphs Schemjakin proceeds from the completely illegal application of quantum mechanics to the diffusion of the substance in solution. We accept Schemiakin's considerations as valuable generalizations of the experimental data. We decisively reject his theory and we shall construct our theory on the base of diffusion and supersaturation" (Zeldovich and Todes 1949, p. 182).

As a matter of fact, the emission-wave theory had been taken under criticism earlier. In 1934, the meeting of Scientific Counsel of Chemical Faculty took place to discuss Mikhalev's Ph.D. thesis. Prof. Academician Ivan Alekseevich Kablukov (1857–1942), who was a referee, pointed to some defects of Mikhalev's text. In particular, he said that the emission-wave theory presented in this thesis had not a sufficient empirical base.

However, Zeldovich and his coauthor's attitude to the emission-wave theory became standard. Referring to Schemjakin's and Mikhalev's book, the 2006 article states: "the quantum-mechanical "explanation" of Liesegang rings as a visualization of de Broglie waves cannot be taken seriously" (Skorobogatov and Kamenskii 2006, p. 826).

References

Dhar NR, Chatterji AC (1925a) Theory of Liesegang rings formation. Kolloid Z 37:2–9

Dhar NR, Chatterji AC (1925b) Theory of Liesegang rings formation. Kolloid Z 37:89–97

Dhar NR, Chatterji AC (1928) Formation of Liesegang rings and peptizing effect of gels. Kolloid Z 40:97–112

Dogadkin B (1926) Periodicheskie sloi osadka fosforno- kaltsievykh solei (The periodic precipitate layers of phosphor-calcium salts). In: Gosydarstvennyi Timeriazev issledovatelskii Institut Stat'i. (Timereazev State Research Institute, Papers). Vologda, pp 41–57

Dogadkin B (1928a) Über sekundäre Periodizität Niederschlagsbildung bei Liesegang'schen Ringen. Kolloid-Z 40:136–140

Dogadkin B (1928b) Ob odnom liubopytnom iavlenii (About an interesting phenomenon). Publisher of Timeriazev, Academy of Agricultural Sciences, Moscow

Dunin MS, Schemjakin FM (1926) Die Bildung des sekundären Systems der Liesegan'schen Schichtungen. Kolloid-Z 39(1):50–53

Dunin MS, Schemjakin FM (1928a) "Razdelenie semian kenafa po udel'nomu vesu" ("Separation of kenaf seeds by specific gravity"). On the farming of kenaf. Bednota (Poor Publisher), Moscow, pp 219–252

Dunin MS, Schemjakin FM (1928b) "Magnetizatsia semian" ("Magnetization of seeds"). Bednota (Poor Publisher), Moscow

Dunin MS, Schemjakin FM (1929) Die Bildung des sekundären Systems der Liesegang'schen Schichtungen. II. Mitteilung. Kolloid-Z 48(2):167–170

Hedges ES (1931) Colloids. Edward Arnold & Co., London, 272P

Hedges ES (1932) Liesegang rings and other periodic structures. Chairman and Hall Ltd., London

Hedges ES, Myers JE (1926) The problem of physico-chemical periodicity. Arnold & Co., London

Henisch HK (1988) Crystals in gels and Liesegang rings. Cambridge University Press, New York

Kingsland SK (1985) Modeling nature. Episodes in the history of population ecology. Chicago University Press, Chicago 266 pp

Küster E (1913) Beiträge zur entwicklungsmechanischen Anatomie der Pflanzen. Heft 1. Jena

Leduc St (1911) The mechanism of life. Redman Co., New York

Liesegang R. (1896a) Über einiger Eigenschaften von Gallerten, Naturw. Wochschr. Bd.11. N 30. S.353–363

Liesegang R (1896b) A-Linien, Photogr Archive 37(N 801), 321–326

Liesegang R (1897) Chemische Vorgange in Gallerten. Photogr Arch 38(Heft 9):129–131

Liesegang R (1907) Über die bei Diffusion auftretenden Schichtungen. Zeitschrift Physik Chem 59(Heft 4):444–447

Liesegang R (1911) Zur Übersättigungstheorie einiger scheinbar rhythmischer Reaktionen. Zeitschrift Physik Chem 75(Heft 3):371–377

Liesegang R (1913) Innere Rhythmen in Pflanzenreich. Naturwissenschaften Wochenschrift. 12 Band. N 25:S.1–10

Liesegang R (1929) Autobiographisches von Raphael Eduard Liesegang. Kolloid Z Bd.49. S.226–667

Mikhalev P, Nikiforov W, Schemjakin FM (1934) Über eine neue Gesetzmäßigkeit für periodische Reaktionen in Gelen. Kolloid-Z 86(N 2):197–200

Mikhalev P, Ulianov A, Schemjakin F (1939) Vlianie radiatsii indutsirovannoi korroziei metalov na destruktsiu periodicheskogo osadka v gele (Influence of the radiation induced by the corrosion of metals on the destruction of the periodical precipitate in gel). Doklady AN SSSR 35(1):33–39

Niederson U, Kuhnert L (ed) (1987) Selbstorganisation chemischer Strukturen. Arbeiten von Friedlieb Ferdinand Runge, Raphael Eduard Liesengang, Boris Pavlovich Belousov, und Anatol Markovich Zhabotinsky. Ostwalds Klassiker der exakten Wissenschaften. Leipzig, Bd. 272

Nikolsky KV (1934) Kvantovaia mekhanika molekuly (Quantum mechanics of a molecule). G.T. T.I, Leningrad, Moscow

Nikolsky KV (1936) Printsipy kvantovoi mekhaniki (Principles of quantum mechanics). Uspekhifizicheskikh nauk 26:50–59 (in Russian, there is an English translation in Physics Uspekhi)

Nikolsky KV (1937) 'Otvet V. A. Foku' (A reply to V. A. Fock). Uspehi fizicheskih nauk (4):555

Oglivie M, Harvey J (eds) (2000) The biographical dictionary of women in science, vol 2. Routledge, NY, London, p 1325

On 60th Birthday of Fedor Mikhailovich Schemjakin (1965) Zhurnal analiticheskoi khimii. 20(6):764

On 70th Birthday of Fedor Mikhailovich Schemjakin (1975) Zavodskaya laboratoria 41(2):536

Ostwald W (1912) Onlines of general chemistry (trans: Taylor W). Macmillan and Co. 1, 596P

Ostwald W (2017) The autobiography (Trans. Jack R). In: Scholz F, Jack R (eds). Springer International Publischer, Berlin

Pechenkin AA (2012) The early statistical interpretations of quantum mechanics in the USA and USSR. In: Studies in the history and philosophy of modern physics. vol 42, pp 25–34

Prager S (1956) Periodic precipitation. J Chem Phys 25:279–283

Raman CV, Ramaiah KS (1939) On the wave like character of periodic precipitates. Proc Indian Acad Sci Math Sci 9(6):455–466

Schemjakin FM (1931) *Provodite opyty po namagnichivaniu semian* (Let us conduct experiments with magnetization of seeds). Moscow, 2nd edn. Gos izd selskokhoziastvennoi literatury, Moscow

Schemjakin FM, Mikhalev PF (1938) Fiziko-khimicheskie periodicheskie process (Physico-chemical periodical processes). Leningrad, Moscow. Izdatelstvo AN SSSR. Bibliography: 925 titles

Schleussner CA (1924) Diffusionsvorgange in Gelatine. II. Kolloid-Z 34:336–341

Servos JW (1990) Physical chemistry from Ostwald to Pauling. The making of a science in America. Princeton University Press, New Jersey. 438 p

Skorobogatov GA, Kamenskii AV (2006) A model of formation of Liesegang Rings under stimulated precipitation conditions. Russ J Phys Chem 80:714–725

Smith D (1984) On Ostwald's supersaturation theory of rhythmic precipitation (Liesegang's rings). J Chem Phys 81:3102. View online http://dx.doi.org/10.1063/1.448012

Stern KH (1955) Bibliography of Liesegang rings. University of Arkansas Editorial Service

Veil S (1934) Les phénomènes périodiques de la chimie. Theil 1: Les périodicities de structure. Theil II. Les periodicities cinetiques. Hermann, Paris

Veil S, Bull L (1931) Microscopic and cinematographic study of Liesegang rings. Compt Rend 192:282–284

Zeldovich IB, Todes OM (1949) O matematicheskoi teorii periodicheskoi sedimentatsii (About a mathematical theory of periodical sedimentation). Zhurnal physicheskoi chimii 23(2):180–191

Zeldovich IB, Barenblatt GT, Salganik RL (1961) O periodicheskom osashdenii pri vzaimnoi diffuzii dvukh veshestv (On quasiperiodical precipitation at the mutual diffusion of two substances (Liesegang rings)). Doklady AN SSSR. 140(6):1281–1284. (English version: Zeldovich YB, Barenblatt GT, Zalganik RL Sov Phys Dokl 6:869 (1962))

Chapter 3
A Research Area

Abstract Chapter 3 describes the development of the research area which can be called "chemical periodical processes". It emphasizes the phenomenon of four books: German 1913 book, English 1926 book, French 1934 book, and Russian (Soviet) 1938 book "Physico-chemical periodical processes". This list can be enriched by including one more book, namely, "Liesegang rings and other periodical structures" (London, 1932). Kremann 1913 book states that the periodical reactions can be divided into two main groups: homogeneous and heterogeneous. However in the succeeding books this classification has been overshadowed by the other problems. Besides Liesegang rings and other periodical structures and processes these books are concerned with mathematical modeling of chemical periodicity. This is a specific subarea of the area of chemical periodicity. Lotka's 1910 paper showed that by taking into consideration the phenomenon of autocatalysis one can propose the system of equations describing the slowly damped chemical oscillations, Lotka's 1920 paper showed that by modifying his 1910 model one can produce the system of equations described (in particular) undamped oscillations. Besides Lotka, Hirniak (Lvov) contributed to the early mathematical studies of chemical periodical processes (1908, 1911).

Keywords Research area · Development of science · Periodic phenomena
Classification · Thermo-kinetic oscillations · Mathematics · Differential equations
Damped and undamped oscillations · Self-oscillations · Autocatalysis

3.1 Introductory Comments

R. Liesegang and his followers stimulated research in chemical periodic processes. A new research area appeared. This area embraced not only the exploration of Liesegang rings and similar structures. It included research in electrochemistry, the dissolution of metals, periodical catalytic reactions. By explaining periodic phenomena, some of the researchers turned to Liesegang rings as analogies. As we have seen, Wilhelm Ostwald recalled his explanation of Liesegang phenomenon,

© The Author(s) 2018
A. Pechenkin, *The History of Research on Chemical Periodic Processes*,
SpringerBriefs in History of Science and Technology,
https://doi.org/10.1007/978-3-319-95108-9_3

when he examined the periodic evolution of hydrogen in the process of dissolution of chromium in the acid. However, many periodic phenomena were described without any impulse from Liesegang and his followers. So, in 1913, Robert Kremann's book about the periodic chemical phenomena has been published. Kremann had not mentioned Liesegang rings, and he wrote about periodical processes on dissolving of metals, periodic phenomena during electrolysis of salt solutions (the release of some element on anode/cathode and its following dissolving), periodic catalysis (some metals might decompose hydrogen peroxide periodically). Kremann wrote about Alfred Lotka's mathematical theory of periodical processes. He also mentioned J. Hirniak's theoretical considerations. He held that Lotka had formulated the kinetics of heterogeneous processes and Hirniak had formulated the kinetics of homogeneous processes. True, the subsequent books on chemical oscillations have not upheld such classification.

The idea of periodicity was in the air in the 1910s. In Germany arose Schwingungslehre (Hort 1910; Barkhausen 1911), which gave impulse for the development of the general theory of oscillations in the USSR. The Soviet chemists referred to dialectics which proclaimed the unity of the world. In the world, winter gives way to summer, and night gives way to day.

3.2   Mathematical Foundations of the Theory of the Periodic Reactions: Lotka, Hirniak

3.2.1   Alfred Lotka

There is Alfred Lotka's (1880–1949) biography in the *Dictionary of Scientific Biography*. Lotka's biography is present in the book on the history of mathematical methods of ecology (Kingsland 1985; Lotka-Volterra Approach… 1985). This writing concentrates on Lotka's achievements in mathematical statistics, ecology, and demography. Here, we concentrate on his contribution to chemical kinetics, the contribution which was made in the early period of his scientific activity.

In 1999, Zhabotinsky wrote that "oscllations in a number of heterogeneous systems were described at the end of the last century. A decisive step was made in Lotka's theoretical papers demonstrating that the concentration oscillations are possible in a simple system following the law of mass action. Besides, Lotka showed the interconnection of the concentration oscillations with autocatalysis" (Zhabotinsky 1991, p. 3).

As was noted above, the concept of autocatalysis was put forward by Wilhelm Ostwald.

In 1910, Lotka published an article about the theory of periodic reactions in *Zeitschrift für physikalische Chemie* published by Wilhelm Ostwald and van't Hoff. There is an English version published in the *Journal of Physical Chemistry*.

Lotka considered the series of chemical reactions each of which is (practically) irreversible (Lotka 1910b):

$$a \rightarrow A$$

$$A + B \rightarrow 2B$$

$$B \rightarrow C$$

Lotka meant that "the capital letters refer to substances in the state of "dilute" gas or solution, while the small letter a denotes a saturated vapor or solution in contact with its condensed phase. It is further to be assumed that the conversion of a into A is slow as compared with the establishment of equilibrium between the condensed phase and its vapor or solution, so the concentration of the latter may be considered as practically constant" (Lotka 1910a, p. 271).

In this connection, Zhabotinsky writes that a denotes a reservoir tank which is practically inexhaustible source of A. Then, the equations expressing the rate of change of the concentrations of the substances A and B are:

$$\frac{dC_A}{dt} = H - k_1 C_A$$
$$\frac{dC_B}{dt} = k_1 C_A - k_2 C_B$$

Let us assume that the substance B influence autocatalytically its own rate of formation, and let us further assume that this influence follows the simplest possible law, so that we can write for k_1 in the first scheme $k_1 = k c_A$

Equations then become

$$\frac{dC_A}{dt} = H - k C_A C_B$$
$$\frac{dC_B}{dt} = k C_A C_B - k_2 C_B$$

After some transformation of the above system, Lotka came to the system describing usual damped oscillations. However, under some conditions expressed mathematically, Lotka's system describes a periodic reaction (damping of oscillations is negligible).

At the end of his paper, Lotka wrote that "no reaction is known which follows the above law. And as a matter of fact the ease here considered of matters lying outside the field of physical chemistry. It is seems interesting, however, also from a purely chemical point of view, to note that in a system in which consecutive reactions take place in the presence of an autocatalytic decomposition product, we gave the requisite conditions for the occurrence of a periodic process" (p. 274).

In 1912, Lotka published the paper which purpose was "to present a general method sufficiently broad to be applicable not only to a large class of cases of interest to the physical chemist (including for example complicated series of consecutive reactions which need not even obey the law of mass action), but also to the cases of biological interest" (Lotka 1912, p. 235).

For the system with two variables, Lotka again found the condition when the damped solution becomes oscillatory (damping is negligible).

Let us turn to Lotka's 1920 papers (Lotka 1920a, b; Kremann's book was published in 1913, Kremann took under consideration Lotka's 1910 and 1912 papers). Zhabotinsky wrote that Lotka's 1920 model "represented the major breakthrough in oscillatory chemistry. It contains two consecutive autocatalytic steps resulting undamped oscillations" (1991, p. 379).

By using Lotka's symbolism in his 1910 paper, one can state the following scheme for his 1920 paper:

$$A + B \rightarrow 2A$$

$$A + B \rightarrow 2B$$

$$B \rightarrow,$$

which corresponds to the system of differential equations describing two consecutive autocatalytic steps.

Lotka's 1920 article is basically dedicated ecological problems (he came to the construction which is known as the Lotka–Volterra equation). By the end of his 1920 paper, Lotka writes that his equations are applicable to the concentration oscillations which chemistry takes into account. Below one finds Lotka's equations in Zhabotinsky's interpretation:

$$\frac{dC_A}{dt} = k_1 C_A - k_2 C_A C_B$$
$$\frac{dC_B}{dt} = k_3 C_A C_B - k_4 C_B$$

Lotka also wrote about universality of the oscillatory processes in nature. However, he did provide again any definite examples of chemical reactions. "In conclusion it may be remarked that a system of equations identical in form with (8), (10) (in our presentation we have the above system) is obtained in the discussion of certain consecutive autocatalytic chemical reactions. Here, however, the coefficients A, B are constants and the integration can be reduced to a quadrature (in our case the coefficients k_1, k_2, k_3, k_4). Aside from a certain number of periodic reactions which have been observed more or less as laboratory curiosities, a certain interest is also attached to this matter from the fact that rhythmical reactions (e.g., heartbeat, which may continue after excision), play an important role in physiology" (Lotka 1920a, p. 415).

3.2.2 Julian Hirniak

In 1910, Hirniak in "Zeitschrift für physikalische Chemie" proposed that cyclic reactions can be oscillatory. He used the simplest example: the cyclic interconversion of three isomers. If the clockwise reactions are relatively rapid and the counterclock ones are relatively slow, it is possible to observe damped oscillations in the system. It seems evident that, if one puts all the molecules in the form x_1, then most of them will be converted to x_2, the to x_3, then back again to x_1, and so on.

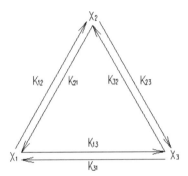

The reciprocal transformation of 3 isomers is considered mathematically by the author as regards the possibility of periodic reactions. Hirniak discussed his equations in their relation to the work of Lotka. He insisted that he was the first who demonstrated the periodic chemical reaction which is based on the laws of chemical kinetics in 1908. However, he published his result in a little-known collection of papers published in Rusyn language by Shevchenko scientific society (Sammerschrigft der mathematisch—naturwissenschaften der Shevchenko Gesellschaft). It should be noted that Lotka in his 1912 and 1920 papers referred to Hirniak's 1908 paper.

There are historical papers about Lotka. Lotka's biography is considered in a number of books. Who was Hirniak? Modern Ukrainian Encyclopedia (vol. 5, 2006, p. 638) provides some information about this scientist. Julian Iosipovich Hirniak (1881–1970) was born in Ternopil region of Ukraine. Before the World War 1, this region belonged to Austria-Hungary Empire. Hirniak was born in small town Strusiv (now Terebovlia, Ternopil region) and died in State New Jersey, USA. In 1905, he graduated from Lvov Higher Technological School and received Ph.D. In 1904–05, he worked for chemical laboratory in Prague. He also worked as a teacher there. He had a kind of fellowship in Leipzig (the Encyclopedia gives no details here). He was a professor at Ukrainian State University in Kamianets-Podilskyi in 1918–1919. Within 1921–1923, he was Professor of Lvov Ukrainian "secrete" university. In 1923, he became rector of the secrete higher technological school in Lvov (here "secrete" means nonofficial). In 1939, he was docent at Higher Polytechnic School. In 1944, he moved to Germany; in 1950, he moved to the USA.

Hirniak was a member of Shevchenko scientific-technological society in Lvov (Ševčenko-Gesellschaft in Lemberg).

Encyclopedia says that the most important Hirniak papers were concerned with physical chemistry. It does not say that the very important Hirnak's paper was published in Rusyn language. However, Hirniak writes himself about it in his fundamental paper "Zur Frage der periodischen Reaktionen" published in *Zeitschrift für physikalische Chemie*. Bd. 75, 1911, p. 675. Hirniak believed that Lotka did not refer to this paper because it was published in a rare periodical in the obscure language.

Probably, Hirniak belonged to Rusyn (Ruthenians), a small ethnic group which belongs neither to Russians, nor to Ukrainians. These people live in East Karpaty and on the West Ukraine. The Big Soviet Encyclopedia (1955) says that as the Ukrainian Soviet Republic was established in 1940, Rusyns became a part of Ukrainians and the nationality "Rusyn" is not applicable. Nevertheless, in Austria-Hungarian Empire Ruthenians enjoyed a cultural autonomy and Shevchenko scientific-technological society published works in Rusyn language.

Let us turn to Hirniak's 1911 paper. A. Zhabotinsky in his historical paper writes: "Thermodynamics, however, puts strong restrictions on the rate constants in this system: the product of the clock-wise constants must be equal to the product of the coun-terclockwise constants. This condition immediately forbids any oscillations in the system. Later, it was shown that it is impossible to have any concentration oscillations in the vicinity of the thermodynamic equilibrium state (here Zhabotinsky refers to a number of the 1947–1970 publications). This thermodynamic analysis made a very strong impression on the majority of chemists, who interpreted it as being valid for all homogeneous closed chemical systems" (Zhabotinsky 1991, p. 379).

3.3 The Phenomenon of Four Books

We referred to four books dedicated to chemical periodical processes above: Kremann (1913) "*Die periodischen Erscheinungen in der Chemie.*" Sammlung. Chemischer Vorträge. Verlag von Ferdinand Enke. 1913. Bd. 19. S. 289–416; Hedges and Myers (1926) "*The problem of physico-chemical periodicity.*" Arnold & Co; Veil (1934), *Les phénomènes périodiques de la chimie.* Theil 1: Les périodicities de structure. II. Les periodicities cinetiques. Paris: Hermann; Schemjakin and Mikhalev (1938) "*Fiziko-khimicheskieperiodicheskie process*" ("*Physico-chemical periodical processes*"). Moscow; Leningrad: Izdatelstvo AN SSSR. 173pp. Bibliography: 925 titles (in Russian).

This list could be extended by mentioning one book more: Hedges (1932), *Liesegang rings and other periodic structures.* London: Chairman and Hall Ltd.

As was mentioned above, Kremann's book does not mention Liesegang rings. It is basically dedicated to the processes rather than the structures. The periodic processes are classified as homogeneous and heterogeneous in Kremann's book.

ЗБІРНИК

МАТЕМАТИЧНО-ПРИРОДОПИСНО-ЛЇКАРСЬКОЇ СЕКЦИЇ

Наукового Товариства імени Шевченка.

ТОМ XII.

ПІД РЕДАКЦИЄЮ

ІВАНА ВЕРХРАТСКОГО, Дра ВОЛОДИМИРА ЛЕВИЦЬКОГО
і Дра СТЕФАНА РУДНИЦЬКОГО

SAMMELSCHRIFT

DER MATHEMATISCH-NATURWISSENSCHAFTLICH-ÄRZTLICHEN SECTION

DER ŠEVČENKO-GESELLSCHAFT DER WISSENSCHAFTEN in LEMBERG.

BAND XII.

REDIGIRT VON

JOHANN WERCHRATSKYJ, Dr. VLADIMIR LEWYCKYJ
u. Dr. STEFAN RUDNYCKYJ.

У ЛЬВОВІ, 1908.

Накладом Наукового Товариства ім. Шевченка.

З друкарні Наукового Товариства імени Шевченка
під зарядом К. Беднарського

The cover of the collected papers of Shevchenko society. Hirniak's 1911 paper about the periodical reactions is published in such an issue

All other books put the Liesegang phenomenon in their centers. Hedges and Myers book points to "static periodicity," "periodic structures," and "periodic processes." However, "static periodicity" the "classic example" of which is "the rule enunciated by Lothar Meyer and by Mendeleev," the rule that "the physical and chemical properties of elements vary periodically with increasing atomic weight" (Hedges and Myers 1926, p. 17), was not present in the books about periodicity published in the 1930s. The essays on physico-chemical periodicity came to concentrate on chemical reactions and chemical structures.

Hedges' book contains the following classification: periodic structures, periodic chemical reactions, miscellaneous periodic reactions. In turn, periodic structures embrace (1) periodic precipitation (the characteristic pattern of which is the Liesegang rings (mentioned in Sect. 3.1), (2) periodic structures in presence of chemical reactions, (3) periodic diffusion without chemical reactions, (4) periodic crystallization and allied phenomena.

Schemjakin and Mikhalev gave another classification. They divided all the periodical processes that were in the focus of their book into three large categories: chemical processes, physico-chemical, and physical.

The chemical processes embrace: (1) periodic precipitation (Liesegang rings), (2) periodic liquid evolution, (3) periodic gas evolution, (4) periodic catalysis, (5) periodic enzymatic processes, (6) periodical electro-chemical processes, (7) periodic photochemical reactions, (8) periodic florescence, (9) periodic gas reactions, (10) periodic corrosion.

According to Schemjakin and Mikhalev, the physico-chemical processes embrace: (1) periodic adsorption, (2) periodic condensation, (3) periodic coagulation, (4) periodic salting out, (5) chemotaxis, (6) periodic phenomena on the surface of the active substances. Schemjakin and Mikhalev explained chemotaxis by referring to the movement of colloidal particles in response to the diffusion of organic compounds into the colloidal solution.

Schemjakin and Mikhalev refer to the periodic crystallization, periodic diffusion, the jumps of drops resulting from syneresis among the physical periodic process.

The phenomenon of four books evidences that a new area of research arose in chemistry. This area was located partially in physic chemistry, partially in colloid chemistry. This was not a scientific discipline: no special university departments, no special textbooks and correspondingly no "ideals of natural order" (the terminology of S. Toulmin's philosophy of science).

A research area presupposes communication between scientists belonging to different scientific establishments and different countries. It presupposes cross-references and an exchange of letters (judging on Schemjakin–Mikhalev's book, Schemjakin and S. Veil wrote to each other, Hedges, as we have seen, took Schemjakin's result under consideration, Schemjakin and Mikhalev cited Hedges and his coauthors' papers). A new research area turned out to be the stimulus of new scientific investigations. In the middle of XXth century, it was enriched by research in thermochemical oscillations, and in the second half of XX the century, it was enriched by the Belousov–Zhabotinsky reaction which became the point of reference for the following studies on chemical oscillations.

3.4 Thermokinetic Oscillations

Thermochemical oscillations have been described by D. A. Frank-Kamenetskii. Basically, he was a specialist in physics of plasma. Since 1946–1956, he was a researcher in Arzamas 16, a researcher in nuclear physics. Since 1956, he worked for Institute of Nuclear Physics (now it is Kurchatov Institute) where run a research group.

In 1939, Frank-Kamenetskii demonstrated that Lotka's 1920 scheme which mentions three subsequent reactions (two of them are autocatalytic) can be applied in the description of the processes which take place as the higher hydrocarbons are oxidized. A bit later, Franck-Kamenetskii conducted research of these processes in the flow reactor under complete stirring of the arriving mixture with the reacting substance (Gerard, Frank-Kamenetskii 1939a, b). He insisted that these oscillations should be treated as chemical oscillations.

"Gervart and me, Frank-Kamenetskii wrote, took the periodical pulsation of cool flame under consideration. We had reached an uninterrupted supply of the benzene/air (benzene/oxygen) mixture. We could observe the process which was periodical over prolonged periods" (Frank-Kamenetskii 1967, pp. 440–441).

In 1947, Frank-Kamenetskii published the book "Diffusion and, heat transfer in chemical kinetics" (the second edition was published in 1967, and the third one was posthumously published in 1987). The final chapter of this book is entitled "Chemical oscillations." In this book, Frank-Kamenetskii formulated the following definition of thermokinetic oscillations: Besides pure kinetic oscillations, there are periodical reactions connected with both the kinetics and the heat evolution and heat removal.

"From the point of view of mechanism, periodic chemical processes can be divided into purely kinetic oscillations linked only with a change of the concentrations of the intermediate reaction products, and thermokinetic oscillations in which along with the change of concentrations, periodic change of the temperature, due the heat evolved in the reaction, also plays a substantial role" (Frank-Kamenetskii 1939a, b, p. 322).

One more scientist should be mentioned in the history of thermokinetic oscillations. This is Izrael Evseevich Salnikov who cooperated with D. A. Frank-Kamenetskii. Salnikov (1914–2001), however, was a representative of another Soviet scientific school, the school of nonlinear oscillations founded by Alexander Alexandrovich Andronov. The Andronov School appeared within framework of radiotechnology and mechanics (1920–1930). This school was based on the qualitative theory of differential equations founded by H. Poincaré. The central concept which Andronov formulated was the concept of self-oscillations, undamped oscillations in a dissipative nonlinear system, which maintained by external non-periodic energy source.

In brief, the story runs as follows. In 1925, Leonid I. Mandelstam, who graduated from Strasbourg University in 1902 and started as a radiophysicist and optician at the Strasbourg Institute of Physics, took a chair of theoretical physics at the Moscow State University (see: Pechenkin 2014). Around him, a group of talented scientists arose. In 1927, Mandelstam set his graduate student A. Andronov the problem of improving a mathematical technique with the help of which radioengineering device (a tube generator) was approximately described. Andronov's work resulted in important conceptual innovations. In his 1928 paper and in his subsequent Ph.D. thesis, a rigorous mathematical theory of the oscillations typical for a tube generator and evident in many customary engines (say, a clock) and in living beings (say, beats of the heart) had been elaborated upon. Andronov turned to the qualitative theory of differential equations developed by Henry Poincaré in another context. This theory yields a rigid (although qualitative) solution of nonlinear differential equations. It provides the methods of what Andronov and his colleges named qualitative integrating. This is methods to determine the states of equilibrium implied by the differential equation, to analyze their stability, to recognize the periodic solutions represented by closed curves, to find out whether they are stable, etc.

Andronov demonstrated the connection between Poincaré's limit cycles and the above oscillatory phenomena (in tube generators, clocks, etc.). He put the general problem of integrating the differential equation:

$$\mathrm{d}^2 x/\mathrm{d}t^2 + \gamma \, \mathrm{d}x/\mathrm{d}t + \omega^2 x = f(x, \mathrm{d}x/\mathrm{d}t),$$

where x is the generalized coordinate (e.g., the current, excursion of a pendulum), t is the time, γ is a damping factor, ω is the eigenfrequency, that is a frequency of that oscillations which would take place if the damping, and the energy source had been eliminated, $f(x, \mathrm{d}x/\mathrm{d}t)$ is a nonlinear function describing how to operate the energy source included into the control system of self-oscillatory design. The limit cycle is a closed trajectory (hence, the trajectory of a periodic solution) such that no trajectory sufficiently near it is also closed. In other words, a limit cycle is an isolated closed trajectory to which all neighboring ones tend in the course of time (at $t \to \infty$ or $t \to -\infty$). This means that every trajectory beginning sufficiently near this cycle either winds itself upon it or unwinds from it. In the former case, we have a stable limit cycle (Fig. 3.1), and in the latter case, we have an unstable limit cycle.

After a number of successful studies (which Andropov partially conducted with another Mandelstam's former student A. Vitt), the concept of self-oscillations gained popularity. This concept turned out to be the conceptual center of Andronov–Vitt–Khaikin's book "The theory of oscillations" (1937) which appeared in three Russian editions and was twice translated into English (1949, 1965). (This book was written by Andronov in cooperation with two other former graduate students of Mandelstam, Vitt, and Khaikin. As Vitt was arrested, his name never appeared in the first edition—[see: Bendrikov and Sidorova (1981), Pechenkin (2014)]. In connection with the concept of self-oscillations, this book represents the other major conceptual innovations of the Mandelstam–Andronov school, namely the concepts of bifurcation and roughness (structural stability). After War II, two books dedicated to elaboration and popularization of

Fig. 3.1 A stable limit cycle

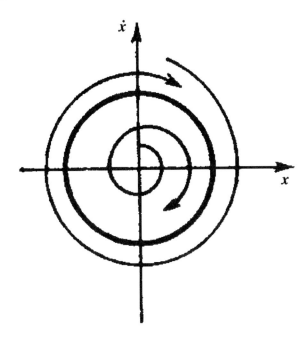

self-oscillations appeared (Teodorchik 1944; Kharkevich 1952). K. Teodorchik's book ran into three editions. "The phenomenon of self-oscillations occurs in nature at every step," G. Gorelik, Mandelstam's former student, wrote in his popular textbook "Oscillations and Waves" (1950).

At the beginning of the 1930s, Andronov moved to Gorky city (now: Nizhny Novgorod) where around him a scientific community has been formed. This community can be considered as a branch of the Mandelstam School. The main concern of Andronov's community was to develop the concept of self-oscillations for multidimensional systems applicable to control engineering and for continuous media. Y. I. Neymark, Andronov's former graduate student, said to the present author [in the following passage, he mentioned his article in the popular scientific journal "Priroda" (Neymark 1991)]:

> I was like a bloodhound on the scent of self-oscillations. Where did I not see self-oscillations? My graduate student Y. I. Gorodetsky was concerned with self-oscillations in metalworking... I dealt with self-oscillations in burning, now I am concerned with – you have seen the journal *Priroda* – self-oscillations in society.

Let us return to chemical oscillations. I. Salnikov, who published an article on homogeneous oscillatory reactions in 1949, was mentioned above. Salnikov was a student of G. Gorelik whose textbook has just been mentioned. His studies in chemical oscillations were initiated by Andronov. In his first article on self-oscillations, Andronov pointed to periodic reactions as an example of self-oscillations (Andronov 1929, p. 561). He referred to R. K. Kremann's "Die periodischen Erscheinungen in der Chemie" which has been cited as one of the "four books." This reference was purely ideological: This book does not show any self-oscillatory chemical reaction, if by self-oscillations we understand phenomena described by the above type of differential equations whose "qualitative integration" yields limit cycles.

Andronov also referred to A. Lotka's writings (Andronov 1929, p. 561; Andronov and Khaikin 1937). Lotka's 1910–1920 articles (see above—Sect. 3.2 of this chapter) are very important for Andronov and his collaborators. However, Lotka had not come to self-oscillatory model of chemical reaction. In his 1910 article, Lotka formulated equations for dissipative systems, the equations describing damping oscillations. He meant the following mechanism:

$$A \xrightarrow{k_0} X \xrightarrow{k_1 \leftarrow - - -} Y \xrightarrow{k_2} B \longrightarrow$$

(where $\xrightarrow{\leftarrow - - -}$ designates self-catalysis (or autocatalysis), that is, $X + Y \to 2Y$),

which corresponds the following system of equations:

$$dx/dt == k_0A - k_1xy$$

$$dy/dt == k_1xy - k_2y$$

As was noted, in 1920 in his article published in *The Journal of American Chemical Society,* he improved his mathematical model. He considered the following chemical reaction:

$$A \xrightarrow{k_0} X \xrightarrow{k_1 \leftarrow - - -} Y \xrightarrow{k_2 \leftarrow - - -} B \longrightarrow$$

And he formulated the following equations:

$$dx/dt == k_0Ax - k_2xy$$

$$dy/dt == k_2xy - k_3y$$

These equations already describe undamped oscillations. But these are not self-oscillations because they are not stable. By referring to Lotka's model as a self-oscillatory model, Andronov suggested that if this model had been improved, it would give self-oscillations.

By setting Salnikov the task to study chemical self-oscillations, Andronov suggested that Salnikov would continue Lotka's work and he would put this work on an experimental base. Salnikov was sent to cooperate with Frank-Kamenetskii at Institute of Chemical Physics. As was noted, Frank-Kamenetskii had already described the oscillatory behavior of cool flames. As Salnikov wrote in his recollections, his aim was to turn Frank-Kamenetskii into a partisan of self-oscillations (Salnikov 1992, p. 8). As a result, Frank-Kamenetskii referred to self-oscillations in his 1942 article. However, this reference was ideological:

Frank-Kamenetskii did not use Andronov's methods, and with the benefit of Andronov's terminology, he interpreted what he obtained by his own methods. Frank-Kamenetskii proceeded from Lotka's 1920 model and additionally assumed the "active by-products" whose concentrations increase according to the law $x = x_0$ $\exp(k_1 a t)$, where t is time, a is the concentration of coreagent, and k_1 is the constant of the rate (and correspondingly for y). By using the Lotka equations, he calculated the critical values of x and y (x_{cr} is the value of x at which dy/dt turns out to be zero, and likewise for y). He showed that at these assumptions the concentrations x and y oscillate around x_{cr} and y_{cr}.

In 1943, in *The Journal of Physical Chemistry,* Frank-Kamenetskii published with Salnikov an article. They tried to solve the problem within the framework of the Andronov paradigm (not only using his ideology). By developing the Lotka equations, they tried to obtain equations describing self-oscillations of cool flame. As Salnikov pointed out later, their attack was not successful enough. Lotka's equations don't fit the process which is not isothermal (Salnikov 1992, p. 10).

In 1949, Salnikov published an article to which we have referred above. He used the idea of thermal catalysis at which he had arrived with Frank-Kamenetskii. This idea is the following: The Arrhenius equation $k = A \exp(-E_{akt}/RT)$, where k is constant of action, E_{akt} is energy of activation, and T is temperature, shows that if a reaction is heat producing, then we have an increase in temperature and hence an increase in the velocity of the reaction. Salnikov formulated his system of equations containing exponents and yielding a limit cycle under some conditions (Salnikov 1949).

Although together with Frank-Kamenetskii Salnikov has published one paper (Frank-Kamenetskii and Salnikov 1943), his research was influenced by his cooperation with Frank-Kamenetskii. Salnikov (1949) formulated a more accurate concept of thermokinetic oscillations by taking into account Andronov's concept of self-oscillations. Salnikov demonstrated that the stable oscillations were possible and formulated the scenarios of their formation.

The experimental research in thermokinetic oscillations took place not only in the USSR; they were conducted in France, England, and USA in the 1960s and 1970s (Field and Burger (eds.) 1985, Chap. 15).

It is interesting that neither Zhabotinsky, nor Scott referred to thermokinetic oscillations in their historical essays (true, Scott mentioned Salnikov's 1949 paper). For these scientists was obvious what is said in the beginning of the present book and what is stated above: the Belousov–Zhabotinsky reaction became the point of reference for the studies in the chemical oscillations in the 1960s and later. The thing is that thermokinetic oscillations are the particular and specific case of chemical oscillations.

Richard Field explains: "Thermokinetic oscillators are a form of emptying/refilling oscillator. In the simplest case, consider a flow reactor in which the exothermic chemical A à X is occurring. A is in the feed stream (rate = kf) of the reactor, and X is an unreactive species that is simply washed out of the reactor. The rate constant for the reaction is given by $k = A \exp(E/RT)$, where E is the activation energy. Heat is lost from the reactor when its temperature is higher than the ambient temperature. If the reactor starts full of A, the reaction begins and heat is produced.

The temperature of the contents of the reactor then begins to rise if heat is produced faster than it is lost to the surrounding. As the temperature increases the reaction rate increases exponentially, the temperature begins to rise very rapidly, and almost all of *A* is consumed. The reaction thus stops; the reactor cools and refills with *A*. Eventually, the concentration of *A* reaches a high enough value in the reactor for the thermally autocatalytic destruction of *A* to explode and again rapidly raise the temperature of the reactor and consume essentially all of the *A*. Again the reactor cools and refills with *A* until another thermal explosion occurs" (Field's Jun 17, 2017 letter to the present author).

References

Andronov A (1929) Les cycles limites de Poincaré et la theorie des auto-entreteneous. C.R. Acad Sci Paris 189(15):559–561

Andronov AA, Chaikin SE (1937) Teoria kolebanii (Theory of oscillations). Gos. Izd, Moscow

Barkhausen H (1950) Einführung in die Schwingungslehre, Dritte Auflage. S. Hirzel Verlag, Leipzig (1 Auflage, 1911)

Bendrikov GA, Sidorova GA (1981) Alexander Adolfovich Vitt, Istoria i metodologia estestvennykh nauk. Izd. MGU im. M.V.Lomonosova, vol 26, pp 150–168

Ebeling W, Peschel M (ed) (1985) Lotka-Volterra approach to cooperation and competition in dynamic systems. Akademie-Verlag, Berlin

Field R, Burger M (eds) (1985) Oscillations and traveling waves in chemical systems. Wiley

Frank-Kamenetskii DA (1939a) Periodicheskie processy v kinetike okislitelnykh reaktsii (The periodical processes in the kinetics of oxidation reaction). Doklady AN SSSR 25:67–69

Frank-Kamenetskii DA (1939b) Diffusion and heat transfer in chemical kinetics. There is an English translation: "Diffusion and heat exchange in chemical kinetics". Transl. from the Russian edition by N. Thon. Princeton University Press, Princeton, 370 pp (1955) (in Russian)

Frank-Kamenetskii DA (1967) Diffusion and heat transfer in chemical kinetics (in Russian). M. izd. Nauka, Moscow (1981)

Frank-Kamenetskii DA, Salnikov IE (1943) About homogeneous oscillatory reaction. Zhurnal fizicheskoi khimii 17(2):79

Gorelik GS (1950) Kolebania i volny (Oscillations and Waves). GITTL, Moscow

Hedges ES (1932) Liesegang rings and other periodic structures. Chapman and Hall Ltd., London

Hedges ES, Myers JE (1926) The problem of physico-chemical periodicity. Arnold & Co

Hirniak J (1911) Zur Frage der periodischen Reaktionen. Zeitschrift für physikalische Chemie. 75:675–680

Hort W (1910) Technische Schwingungslehre, 2nd edn. Julius Springer, Berlin

Kharkevich A (1952) Avtokolebania (self-oscillations). Gostekhizdat, Moscow

Kingsland SK (1985) Modeling nature. Episodes in the history of population ecology. Chicago University Press, 266 pp

Kremann R (1913) Die periodischen Erscheinungen in der Chemie. Sammlung. Chemischer Vorträge. Verlag von Ferdinand Enke. Bd.19:289–416

Lotka A (1910a) Contribution to the theory of periodic reactions. J Phys Chem 14:271–274

Lotka (1910b) Zur Theorie der periodischen Reaktionen. Zeitschrift für physikalische Chemie. 72:508–511

Lotka A (1912) Change of state. Discussion of a general case. Phys Rev 34:235–238

Lotka (1920a) Analitical note on certain rhythmic relations in organic systems. Proc Natl Acad Sci USA 6:410–415

Lotka A (1920b) Undamped oscillations derived from the law of mass action. J Am Chem Soc 42:1595

Neymark Y (1991) Prostye matematixheskie modeli (Simple mathematical models). Priroda 11:9–18

Pechenkin A (2014) L.I. Mandelstam: research, teaching and life. Springer, Heidelberg

Salnikov IE (1949) On the theory of periodic homogeneous reactions, part II. Zhurnal fizicheskoi khimii 23:258

Salnikov IE (1992) U istokov teorii khimichaskih avtokolebanii (At the beginning of the theory of chemical self-osciillations). Dynamics of Systems. Dynamic and Optimization. Termokineticheskie kolebania. Nizhnyi Novgorod: 3–24

Schemjakin FM, Mikhalev PF (1938) Fiziko-khimicheskie periodicheskie process (Physico-chemical periodical processes). Leningrad, Moscow. Izdatelstvo AN SSSR. Bibliography: 925 titles

Teodorchik K (1944) Avtokolebatelnye sistemy (Self-oscillatory systems). Moscow

Veil S (1934) Les phénomènes périodiques de la chimie. Theil 1: Les périodicities de structure. Theil II. Les periodicities cinetiques. Hermann, Paris

Zhabotinsky A (1991) A history of chemical oscillations and waves. Chaos 1:379

Chapter 4
The Belousov Reaction

Abstract The modern history of research in chemical periodical processes started in 1951 when B. P. Belousov discovered the concentration oscillations between oxidized and reduced forms of cerium in the process of the oxidation of citric acid by bromate. Chapter 4 provides Belousov's biography and contains an attempt to understand what kind of man he was. By basing of his publications and notes we try to reconstruct the presumptions of Belousov's discovery proceeding from his research work for Institute of Biophysics of the Academy of Medical Sciences. Chapter 4 provides an extensive use of the recollections of Belousov's contemporaries and interviews taken by the author. By applying T. Kuhn's technique of paradigms we want to explain why Belousov's 1951 and 1955 papers have not been published. We conclude that Kuhn's paradigms provide us with a partial explanation only. A number of social and psychological factors which are not embraced by T. Kuhn's technique is essential.

Keywords Russian revolution · Civil war · Chemical defense
Nuclear weapon · Penetrating radiation · Protectors · Krebs cycle
Dialectics · Scientific publication · Bureaucracy · Academicians

4.1 Introductory Comments

In the beginning of our text, the book (collected papers) ed. by Field and Burger (1985) has been cited. This book says: "The modern history of research in chemical oscillators started in the USSR in 1951 when B. P. Belousov discovered the concentration oscillations between oxidized and reduced forms of cerium in the process of the oxidation of citric acid by bromate."

Belgium physicist the Nobel Prize winner I. R. Prigogine writes that "in experimental studies, the Belousov–Zhabotinsky reaction plays the same role as does the Brusselator in theoretical studies" (1980, p. 121). The Brusselator is a key

---

Sections 2–6 of this chapter closely follow the article Pechenkin (2009).

© The Author(s) 2018
A. Pechenkin, *The History of Research on Chemical Periodic Processes*,
SpringerBriefs in History of Science and Technology,
https://doi.org/10.1007/978-3-319-95108-9_4

mathematical model for Prigogine's nonlinear thermodynamics, specifically a characteristic representation of the phenomena of self-organization. According to circumstances, a wide range of phenomena of self-organization have been observed in the Belousov–Zhabotinsky systems, for example, the rise of oscillations for a period of the order of a minute and wavelike activity.

H. Hacken also cites the Belousov–Zhabotinsky reaction as a basic model of his synergetics (1977, p. 9).

The dramatic history of this reaction has been incorporated into scientific folklore and is even outlined in a textbook on nonlinear dynamics (Strogatz 1994, pp. 254–255). Several essays on this history have been published. In 1984, V. Poleshuk, a historian and a journalist, published an essay on B. P. Belousov's discovery in the famous Soviet literary journal *Novy Mir* (in which Solzhenitsyn published his first writings). S. E. Shnol', the university research teacher of A. M. Zhabotinsky, wrote two reminiscent recollections concerning Belousov's discovery of the reaction for which he subsequently became famous and about the early stage in Zhabotinsky's further development of that work (Shnol' 1997, 2001). The prominent biologist A. Winfree wrote an essay on the Belousov discovery basing on his rich scientific experience and on his conversations with Russian colleagues (Winfree 1987, pp. 661–663). Finally, Zhabotinsky himself wrote a couple of papers on how he elaborated Belousov's discovery, first under S. E. Shnol' and a bit later independently (Zhabotinsky 1985, 1991). However, here one finds a description of this topic from the point of view of a professional historian.

In the present section, we concentrate on the early events connected with Belousov's discovery. The essays mentioned above report that Belousov's original paper, in which he described discovery of a homogeneous oscillatory reaction and proposed a tentative mechanism for that process, was rejected by two main Soviet chemical journals (*Zhurnal obshchey khimii* <The journal of general chemistry> and *Kinetika i kataliz* <Kinetics and catalysis>) in 1951 and in 1955, respectively. Belousov finally managed to publish a brief abstract in the obscure proceedings issued by the institute where he was employed as head of a laboratory (Institute of Biophysics at the Ministry of Public Health). Belousov died in 1970. His full paper was only posthumously published in 1981. An English translation of the 1951 version of his paper appeared in 1985 (Field and Burger (eds.)). For German translation, see: (Niederson and Kuhnert 1987).

It is interesting that it is not possible to blame external forces for this rejection. Neither the Communist Party authorities nor the state ideologists intervened. The scientific community as represented by the editors of scientific journals and their referees refused to accept Belousov's discovery.

In order to explain the 1951–1955 situation, Arthur Winfree (in the essay mentioned above) points to "the resistance of human nature to observations that do not fit into existing theory" (Winfree 1987, p. 661). S. E. Shnol' displays a similar attitude. "The main obstacle proceeded from textbooks on equilibrium thermodynamics. A well-educated person could not imagine a macroscopic order in chaotic motion of the tremendous amount of molecules. He could not believe that the molecules could altogether come to the one state and then to the other state" (Shnol' 1997, pp. 151–152). However, he also places emphasis on a moral aspect of the situation: authoritative Soviet chemists did not want to concern themselves with facts that were provided by an experimentalist engaged in applied research. Poleshchuk also emphasized the moral aspect. He referred to Belousov's inscription on a book given by him as a gift to his lifelong friend: "Don't be shame of not knowing, be ashamed of not willing to know" (Poleshuk 1984, p. 201). Although this inscription was not directly connected with Belousov's unsuccessful struggle for publishing his results, Poleshchuk used it to evaluate the editor's attitude to Belousov's paper.

Let us pay attention to some historical details. The situation with the Belousov's discovery was more complicated than the articles mentioned above might suggest. As was noted above, in 1949, that is, two years earlier than Belousov made his first attempt to publish his results, *Zhurnal Fizicheskoi Khimii* (Journal of Physical Chemistry) published an article concerning oscillatory homogeneous reactions. This was I. E. Salnikov's article that summed up a series of his studies conducted together with David A. Frank-Kamenetskii at the Institute of Chemical Physics since 1941. Frank-Kamenetskii had started to study chemical oscillations at the end of the 1930s. In 1940, he published an article on the oscillations of cool flames—he was the first who obtained these oscillations for a closed system. In 1947, Frank-Kamenetskii published a book which ran into three editions. This book contains a chapter "Periodic processes in chemical kinetics."

It should also be mentioned that in 1957 Shnol' started to publish his results concerning the oscillations of ATPase activity of actomyosin (a kind of protein). Shnol' told to the present author that he experienced difficulties as he attempted to publish his results in leading journals of the Academy of Sciences. Nevertheless, his articles had appeared and, as it was noted above, Shnol' suggested that his student Zhabotinsky to continue Belousov's research as his own dissertation problem. This was at the Physics Faculty of the Lomonosov Moscow State University in 1961. As is well known, it was required that themes of dissertations should be approved by the Chair and by the Scientific Counsel of the Faculty (Department). Therefore, at the beginning of the 1960s, the Belousov discovery was approved by those faculties. Zhabotinsky successfully conducted his research which soon became independent and resulted in a series of publications in the most authoritative Soviet scientific journals and led him to the Lenin Prize (the highest Soviet scientific honor) which he received in 1981. Belousov's name was added to the list of nominees through Shnol's efforts.

At the end of the 1950s, D. S. Chernyavsky, who worked at the Physics Institute of the Academy of Sciences, published a number of articles on periodic processes in photosynthesis (see: Sect. 4.1). So, at the beginning of the 1960s, one can observe a "scientific movement" toward legitimization of chemical oscillations. Articles about chemical oscillations appeared a bit earlier than Belousov's unsuccessful attempt to publish his results, and they appeared a bit later. Why was Belousov's paper rejected? And conversely, why Belousov's contemporaries succeed in publishing their results on chemical oscillations?

4.2   The Belousov Discovery

In his two-page 1959 paper, Belousov described a homogeneous reaction connected with a periodic change in color of an entire reaction mixture: from colorless to yellow, then back to colorless, and so on. This reaction was the oxidation of citric acid by bromate ion, a well-known oxidant. As Belousov explained, the oxidation reaction of citric acid by bromate is usually very slow. The rate of oxidation is increased if cerium cation is used as a transmitter of oxidation. Bromate ion oxidizes Ce^{3+} to Ce^{4+}; in turn, Ce^{4+} oxidizes citric acid, and cerium is reduced to Ce^{3+}. Since the solution containing Ce^{3+} is colorless and the one containing Ce^{4+} is yellow, the reaction undergoes a periodic change in color.

Belousov describes the mechanism of his reaction as follows: First of all, the oxidation of citric acid by quadrivalent cerium takes place. This reaction results in the disappearance of the yellow color and the oxidation of Ce^{3+}. The next stage is the oxidation of Ce^{3+} by bromate to Ce^{4+}. The oxidation of citric acid by Ce^{4+} occurs slowly, and the oxidation of Ce^{3+}, which is a product of the first reaction, proceeds even more slowly. As a result, newly formed quadrivalent cerium could be re-reduced in the former reaction. Quadrivalent cerium does not influence the color of the reaction mixture. However, the oxidation of Ce^{3+} to Ce^{4+} becomes rapid in the course of time. The reason is that this reaction produces bromide, which accumulates in the reaction mixture and reacts rapidly with bromate. The ensuing bromine-producing reaction proceeds even faster. Bromine is immediately captured by hydroxyl acid (acetonedicarboxylic acid) that comes from the oxidation of citric acid by Ce^{4+}. As a result, inactive pentabromoacetone is formed. The removal of bromine leads to the acceleration of the reaction, which forms quadrivalent cerium.

Belousov outlined the following chain of reactions:

$$C_6O_7H_8 \text{ (citric . acid)} + Ce^{4+} \rightarrow C_5O_5H_6 + Ce^{3+} + CO_2 + H_2O$$
$$BrO_3^- + Ce^{3+} \rightarrow^{H^+} Ce^{4+} + Br^-$$
$$Br^- + 2H^+ + BrO^{-3} \rightarrow HBrO + HBrO_2$$
$$H^+ + Br^- + HBrO \rightarrow 2Br^- + H_2O$$
$$3H^+ + 3Br^- + HBrO_2 \rightarrow 2Br_2 + 2H_2O$$
$$C_5O_5H_6 \text{ (acetonedicarboxylic acid)}$$
$$+ 5Br_2 \rightarrow C_3OHBr_5 \text{ (penta-bromoacetone)} + 5Br^- + 2CO_2 + 5H^+.$$

After the disappearance of Ce^{3+}, the color of the mixture is determined by Ce^{4+}. As a result, the mixture is yellow again.

Belousov also assumed the release of free bromine. He thought that acetonedicarboxylic acid is quickly consumed because the rate of its production is very slow. As a result, the liberation of free bromine takes place, and the color of the mixture suddenly becomes yellow. Then, bromine is consumed "gradually but at a definite rate" to form Ce^{4+}.

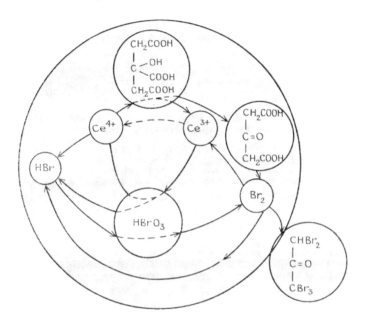

Belousov's scheme of his reaction

As a result of the above processes, the reaction mixture consists of citric acid and bromate, Ce^{4+}, and inactive pentabromoacetone. The above processes will start again. The reaction will stop when "one of the principal ingredients of the reaction mixture, that is citric acid or bromate, is spent" (Belousov 1959, p. 145).

In his early writings, Zhabotinsky (1964) showed that the liberation of free bromine was not essential for the Belousov reaction.

"The reaction described here is remarkable because during its occurrence there is a complex ordered sequence of oxidation–reduction processes, one which is periodically revealed by a temporary change in the color of the entire reaction mixture" (Belousov 1959: 145).

4.3   Biography of Boris Pavlovich Belousov

In his autobiography dated 1950, Belousov has written:

I was born in 1893. I went to the Commerce School in Moscow. In 1908 I left Moscow for Zurich because of health reasons. In Zurich I continued my education first at Realgymnasia and then at the University, from which I graduated in 1915. In 1916 I returned to Moscow.

My first employer was the chemical laboratory at the Moscow Metal Works, where I became an assistant. Then I became Head of the Laboratory of Metallography. From 1919 on I taught at the People's University, in high school, and at the Workers' Faculty. For many years I taught a course of general chemistry at the Higher Military Chemical School of the Red Army. At the same time I taught a course of general chemistry and a course of chemical warfare at the School for Advanced Training of Officers and conducted scientific research.

In the commendation from the Minister of Defense I was thanked for participating in the development of new special courses for the Higher Military Chemical School and for effective scientific research.

In 1935 I left the staff of the Red Army and became a researcher at the All-Union Institute of Sanitation and Chemistry. I totally devoted myself to scientific research.

In the first year of my work for this institution I received an award and was commended for progress in analytical chemistry (detection).

I subsequently worked on the development of stationary automatic gas detectors and the synthesis of medicines and prophylactic means.

In this connection I proposed a new theory of intoxication resulting from vesicant action which was confirmed by other researchers.

Boris Pavlovich Belousov (about 1950). This portrait is placed in the 1985 book "Oscillations and travelling waves in chemical systems"

During Patriotic War I invented a number of valuable medicines (VIP-17, VIP-21, etc.), which were used by the Red Army and were produced by the Acrichine plant.5 In 1950, in connection with work on the removal of some toxic substances from the human organism, I showed the importance of certain biochemical cycles. This should be taken as a basis for the treatment and prophylactics of diseases caused by these substances. That year I also supervised research on antipollution remedies and conducted some other research.

I have repeatedly received awards and commendations from the Ministry of Public Health, Military Authorities, and the Government for my productive and successful work. In particular, I have been awarded Znak Otlicnika (the Order of Excellent Worker), Znak Pocheta (the Order of the Badge of Honor), and the Voroshilov Prize. I have 16 certificates of authorship and 50 scientific papers, more than half of them positively evaluated. I have written a textbook on inorganic chemistry and a number of educational papers as well.

Boris Pavlovich Belousov (the 1960s). This portrait is included into the gallery of the honorable scientists who worked for Institute of Biophysics

Referring to the recollections of Belousov's niece, Mobbi Alexandrovna Belousova, Shnol (1997, p. 146) has written that the young Boris Belousov was close to Marxist intellectuals who in some way participated in the preparation of the October Revolution of 1917. However, as Belousov stated in his personal records (1940, 1950), he was not a member of the Communist Party. According to the material held in the Russian State Military Archives, Belousov started to teach chemistry at the advanced training courses for the officers of the Workers' and Peasants' Red Army (the full official name of the Soviet Army) in 1924. In 1925, these courses became a part of the Military Technological Academy of the Red Army. In 1932, the Military Chemical Academy of the Red Army was formed by the merger of the Chemical Faculty Military Technological Academy and the Moscow Second Institute of Chemistry and Technology. The textbook Belousov mentioned in his autobiography was on inorganic chemistry; he wrote it in coauthorship for military schools (Belousov et al. 1932).

As a researcher at the All-Union Institute of Sanitation and Chemistry, Belousov issued the following reports: "Studies in Dosed Micro Amounts of Chlorine" (1936), "Research in the Field of Detectors of Mustard Gas and Lewisite" (1935), and "Studies in the Reaction of $\beta\beta'$ Diclordiethyl Sulphide With xxx" (1939). Secret substances were indicated as "xxx" on his list of scientific results. In his autobiography, Belousov did not mention that in 1946 he was awarded the Order of Lenin, one of the highest Soviet orders (only the Medal of Hero was higher in the rank of awards). He received this award for his contribution to remedies against phosphorus burns. This research was secret and could not be mentioned in public documents. Belousov did mention his Order of Lenin in his papers of the 1960s.

In 1941 Belousov issued a paper entitled "A New Chemical for the Deactivation of Phosphorus," and in 1942, he issued one entitled "New Chemicals for the Deactivation of Phosphorus and Light Metals." In the summary of the former paper he wrote, "A new remedy (VIP-17) containing a compound xxx with an admixture of catalyst has been worked up. This substance rapidly and completely interacts with phosphorus and its solutions. The application of gauze tampons impregnated by salts with an admixture of catalyst and hygroscopic compounds was proposed (VIP-19)."

In 1952, the new Institute of Biophysics was established as a merger of the All-Union Institute of Sanitation and Chemistry and the Radiation Laboratory of the Ministry of Public Health. This event is not represented in Belousov's 1950 autobiography. This Institute of Biophysics should not be confused with the Institute of Biophysics belonging to the Soviet Academy of Sciences. The latter was also established in Moscow in the early 1950s. In the early 1960s, it was moved to the small city Pushino-na-Oke in the Moscow region. Zhabotinsky was employed there after graduating from Lomonosov Moscow State University.

Belousov's first position was that of a senior researcher (1935). After World War II, he became a laboratory head. In a 1948 memo issued by the director of the Institute of Pathology and Therapy of Intoxications, it was pointed out that his case was exceptional. Belousov did not have any scientific degree and therefore could not technically take the position of the head of a laboratory. His personal file does not contain a diploma from Zurich Technical University; it has only a paper of the Ministry of Public Health that confirms that he graduated from this university.

How did things develop after 1950? At the beginning of the 1950s, the problem of radiation protection became one of the main research problems for the Institute of Biophysics.

The Soviet Union began to prepare itself for possible nuclear war. Belousov was charged with the problem of radioprotection, and later his concern extended to the problem of excretion of radionuclides from a human being. True, Belousov and his colleagues did not speak about "radionuclides"; they used the term "radioactive heavy metals" instead.

Belousov's endeavors in the area of radioprotection are partially represented in the 1963 volume *The Radioprotective Action of Cyanic Compounds* (Rogozkin et al. 1963). The basic idea is expressed thus: "The chemicals which are able to suppress the oxidation processes in an organism immediately before irradiation help the organism to resist the radiation injury" (p. 10).

The effect was related to the ability of anion CN^+ to form temporally stable complexes with a number of respiratory ferments containing iron, resulting in the respiratory depression and the depression of metabolism. Now, it is known that cyanides tie cytochrome system what leads to tissue hypoxia. In particular, they can bind cytochrome a3 and inhibit the activity of cytochrome oxidase, an enzyme responsible for the final step of the electron transport chain from cytochromes to oxygen. … It should be taken into account that cyanides also depressively influence other ferment systems (Rogozkin et al. 1963, p. 99). Belousov wrote chemical and biochemical chapters, in particular a chapter about amygdalin (mandelonitril-beta-gentobiozide), a chemical containing a cyanogen group.

According to reminiscences by Dr. Lev M. Rozhdestvensky and Dr. Gennadi I. Shaposchnikov, Belousov's main idea was to use chitosan as a radioprotector. Chitosan is a linear polysaccharide that is found in the exoskeleton of crustaceans and insect cuticles. Dr. Rozhdestvensky, who collaborated with Belousov on the chitosan project at the beginning of the 1960s and still worked for the Institute of Biophysics in 2005, said that it was an important step in the development of radioprotectors. Although chitosan did not justify hopes, it offered a new methodology for radioprotection. In the 1950s, low molecular sulfhydryl-containing compounds such as cystamine and cysteamine were commonly used as radioprotectors. The efficiency of protection was low, and it required prior injection at high doses that resulted in deleterious side effects, including hepatotoxicity. In contrast, chitosan was a high molecular compound, and its action consisted basically of activating host defenses. Like the generation of radioprotectors that follows (e.g., interleukin-1), chitosan induces the production of endogenous substances that are able to pass ahead of the harmful action of penetrating radiation. Although Belousov never published papers on chitosan, he ran the chitosan project (this protector was named RS-10), and according to some of reminiscences, his retirement in 1966 was due to the failure of the chitosan project.

Belousov, who was a chemist by training, went deeper into advanced biochemical problems. As a matter of fact, he had approached biochemistry earlier, when he had conducted research in chemical warfare. In his autobiography, he mentions his theory of intoxication, but this theory is not represented in the available documents. One can judge Belousov's approach to toxicology by his unpublished manuscript, "On Chemical Presumptions Concerning the Foundations of the Effective Action of the Remedies That Are Tested in the Therapy of Radiation Damage." In this manuscript, he referred to his hypothesis, based on his experience with chemical warfare, in particular with organophosphate agents. The hypothesis, which is apparently out of date now, proposed a chain of biochemical processes. Belousov emphasized that under radiation, cell lecithin decomposes with the release of acetylcholine. In turn, "an accumulation of acetylcholine stimulates the autolysis of tissue proteins," yielding amino acids, in particular histidine.

Under the influence of oxidants, histidine can change into toxic compounds (aldehydes). The manuscript "On Chemical Presumptions Concerning the Foundations of the Effective Action of the Remedies That Are Tested in the Therapy of Radiation Damage" is not dated.

However, judging from the references it can be related to the early 1950s.

Belousov's collaborator Alexei Petrovich Safronov referred to Belousov's hypothesis in his paper entitled "On the Mechanism of Excretion of the Incorporated Polonium by Sodium Diethyldithiocarbamate." With the help of Belousov's hypothesis, Safronov (1961, p. 97) explained the action of his drug.

Evidently, Belousov ran the several central projects at the Institute of Biophysics. According to the present author's interviews, Belousov was a respectable figure at the institute. He had a special, "personal" higher salary, and a car was sent to pick him up to his laboratory and back home (which was important given the distant location of the Institute of Biophysics).

According to Shaposchnikov's recollection, Belousov was familiar with cultural events and sometimes spoke about art with friends. His wife was an actor, but they divorced. Like the majority of his contemporaries, he avoided any political discussion. Prof. Boris Borisovich Moroz told me the following story: In the early 1960s, Prof. Ivan Alexandovich Pigalev (the head of the Laboratory of Therapy), Belousov, and Moroz (then a young researcher) had a tea break at Pigalev's laboratory. Pigalev started to recall the events of the Civil War.

Belousov stood up, said goodbye, and left the laboratory.

His reply to Shnol, who proposed to collaborate with him in order to further study his reaction, was also remarkable (Shnol 1997, p. 160). Belousov said, "All my friends were killed or are dead. I don't want to make new friends." In 1966, Belousov retired (according to some evidence, he was forced to retire). The position of consultant was kept for him, but he never came to the institute again. He died in 1970.

4.4 Oscillations in Analytical Chemistry

Did Belousov's applied research contain analogies to the processes that form his reaction? Did it provide grounds for the construction of such an analogy? The following discussion of this question should be treated as tentative, as it is based on a restricted body of historical documents only.

Although after World War II Belousov concentrated on radiation toxicology, he continued his research in analytical chemistry. Quadrivalent cerium, $Ce(SO_4)_2$, was at hand in Belousov's laboratory, literally and figuratively. Cerate oxidimetry was broadly used in chemical analysis, in particular for oxidimetric titration of uranium (4+). This procedure is based on the equation $2Ce^{4+} + U^{4+} \rightarrow 2Ce^{3+} + U^{6+}$, where Ce^{4+} is converted into Ce^{3+}, as in the Belousov reaction.

Thus, for Belousov, it was a routine chemical reagent that he could take off the shelf at any time.

Bromates and bromate–bromide reagent were widely used in the chemical analysis of organic compounds. Basically, bromate–bromide reagent generated free bromine, which was used, say, to determine the bromine number of hydrocarbons.

Belousov's collaborator Safronov (1959, p. 152) reported about an apparatus constructed by him for the semiautomatic determination of citric acid. This apparatus used bromide. As in Belousov's reaction, it generated bromine, which reacted with citric acid, and this reaction formed pentabromoacetone.

Safronov emphasized that his apparatus should help research on the distribution of radioactive substances in organisms. On the issue on which Belousov's paper on the periodic reaction was published, he and Safronov also published a short paper entitled "New Approaches to the Qualitative Analysis of Cations From the Point of View of an Adopted Theory of Chromaticity." This is a theoretical paper, probably the only one written by them. However, it can be traced back to Belousov's research in color reactions applied to the determination of mustard gas, lewisite, and other war poison substances.

For example, in 1939 Belousov reported a color reaction for β-chlorovinyldichloroarsine with the formation of molybdenum blue. In 1940, Belousov reported the reaction for chloropicrin with the formation of dyes.

The authors wrote: "In the course of our search for reagents for the qualitative determination of uranyl ions we considered the analytical color reaction of a number of heavy metals with organic and inorganic compounds.

By reviewing the relevant literature, we came to the opinion about the essence of the processes resulting in the formation of coloration in some color reactions. We also proposed a classification of some analytical reagents according to the types of their interaction with the ions that should be specified.

Our classification is based on the principle of the intermolecular reversibility of the charge transfer between the atoms of metals that occur in the higher and lower states of valence in a molecule. This principle is common to the concept of chromaticity and embraces inorganic and organic reactions.

Table 4.1 Table illustrating the theory proposed by Belousov and Safronov, taken from Belousov and Safronov (1959); only the three upper rows are reproduced

Name of colored composition of a compound	Formulae	Color of oxides included in the composition of a "mineral blue"
Tungsten blue	$4 \leftrightarrow 6$ WO_2 WO_3	Gray and lemon yellow
Molibdenum blue	$5 \leftrightarrow 6$ Mo_2O_5 $(MoO_3)_3$	Black–violet and white
Vanadium blue	$3 \leftrightarrow 5$ V_2O_3 V_2O_5	Black and orange–yellow

The "mineral blues" provide the simplest example. Their chromaticity is explained by isoatomic charge transfer" (Belousov and Safronov 1959, p. 147).

Belousov and Safronov illustrated their consideration by a table (Table 4.1). They also wrote about "heteroatomic charge transfer," a phenomenon that, according to them, explains the color of substances containing atoms of the different chemical elements that are potentially responsible for the color.

An example is cerium blue $(CeO_2)UO_2$. At the end of their paper, Belousov and Safronov called the charge transfers "oscillations" and wrote about the "isoatomic and heteroatomic oscillations." The Belousov–Safronov conception of chromaticity is rather speculative and does not fit the standards of quantum chemistry. However, there is an excuse. Belousov and Safronov came to an area that was clouded in fog at their time. In the authoritative 1967 volume *Pigments: An Introduction to Their Physical Chemistry*, there is the section "Colour Due to Electron Transfer in Compounds With Metal Ions in Two Different States of Valence" that indicates the problem: "charge transfer of this kind has not yet been investigated very extensively and a general theory is still lacking" (Patterson 1967, p. 26).

In the context of the present investigation, the 1959 paper of Belousov and Safronov provided a model—or rather a metaphor—leading to the idea of Belousov's reaction. We noted that this paper sums up his experience with color reactions in analytical chemistry. Most importantly, a manuscript with a similar title, dated 1947, was included in the list of Belousov's scientific results. Apparently his 1959 paper with Safronov was an outcome of their project, which was launched in the immediate afterwar period. Therefore, it is plausible to assume that this project had stimulated Belousov's meditations, which eventually led him to his oscillatory reaction. As we know, Belousov first submitted his paper about the oscillatory reaction in 1951.

In contrast to the paper of Belousov (1951), the 1959 paper of Belousov and Safronov does not state the oscillations of color and hence does not mention the oscillations in the composition of a reaction mixture. Their paper discusses the electron oscillations that explain the color. Belousov's 1951 paper concerned oscillations in the course of the reaction; the cerium atom temporarily reaches the state Ce^{3+} and then returns to the state Ce^{4+}; then, the process is reiterated. In their paper, Belousov and Safronov considered transitions of electrons from one atom to another within a chemical compound, say, $Mo_2 O_5 \leftrightarrow (MoO_3)_3$. Belousov's reaction is the oxidation–reduction reaction, where there is no electron exchange between cerium and its coreagents. Nevertheless, on a more abstract level, there is a similarity between both papers. Both consider oscillation, the chemical elements in two different states of valence participating in both oscillations. Here, the fog surrounding the problem of the explanation of color on the basis of the charge transfer turned out to be profitable. Belousov and Safronow wrote about the color of chemical individuals. Let us look at the right column of Table 4.1. The two colors are indicated for one chemical individual in it: black–violet and white for molybdenum blue, and so on. This should mean that under the name of molybdenum blue there are two or several compounds containing molybdenum in two valence states. But Belousov and Safronov wrote about the "color of oxides included in the composition of a 'mineral blue.'" This may be understood as follows: Potentially, a chemical individuality such as molybdenum blue contains "two substances" in its composition—black–violet $Mo_2 O_5$ and white $(MoO_3)_3$. As in Belousov's reactions, these "two substances" are connected by electron oscillations.

It is difficult to say whether such a vague analogy could have really led Belousov to his reaction. However, it is clear that analytical chemistry was an area in which Belousov conducted intensive research that influenced his "personal knowledge." The following section attempts another reconstruction of Belousov's discovery.

4.5   From Radiation Toxicology to Periodic Reaction

The manifest thrust came from the idea of the tricarboxylic acid (citric acid) cycle. In his autobiography, Belousov wrote that in connection with his work on the removal of toxic substances from the human organism, he showed the importance of some biochemical cycles. He mentioned the tricarboxylic acid cycle in the aforementioned manuscript too. One may say that the tricarboxylic acid cycle was Belousov's principal concern, as he tried to understand the clinical course of radiation sickness and elaborated the removal of radionuclides from a human organism.

In his posthumously published paper, Belousov (1985, pp. 575–607) himself wrote that the "peculiar behavior of citric acid in relation to some oxidants lies at the foundation of the periodic reaction." He characterized his reaction as a "cycle," which is remarkable because in biochemistry books the tricarboxylic acid cycle is usually pictured as a circular pattern. Arthur Winfree wrote:

"His [Belousov's] interest included biochemistry, and 1950 found him endeavoring to model catalysis in the Krebs cycle using the metal ion cerium instead of protein-bound metal ion common in the enzymes of living cells. The Krebs cycle is a universal part of metabolism by which acetyl residues are oxidized to carbon dioxide in mitochondria.

It is called a "cycle" not because it oscillates in time, but just because the reaction sequence leads in a circle, much as any geochemical cycle. Much to Belousov's surprise, his test-tube caricature, a solution of citric in water with acidified bromate as oxidant and yellow ceric ion as catalyst, turned colorless and returned to yellow periodically for as long as an hour (at room temperature) while effervescing carbon dioxide!" (Winfree 1984, p. 661).

Winfree's description of Belousov's discovery was reproduced in a number of books (e.g., Coveney and Highfield 1995, p. 175).

Biochemical cycles were studied intensively before World War II, culminating in Hans Krebs's 1937 description of the tricarboxylic acid cycle. After World War II, the idea of biochemical cycles found its way into biochemistry textbooks.

Thus, in the 1951 Soviet textbook for medical colleges, Zbarsky et al. (1951) described the tricarboxylic acid cycle but not in its modern form.

Winfree referred to catalytic reactions that were outlined by Albert Szent-Györgyi and other biochemists as early as 1935 and that subsequently entered the textbooks. They were included in the network of metabolic pathways and studied by Hans Krebs in connection with the research that led to the tricarboxylic acid cycle ("Krebs cycle"). These reactions provided the proton transfer suggested by the Krebs cycle. The catalyst was typically an iron ion bound to the porphyrin ring of the heme group of oxidation enzymes. The mechanism may be represented as follows:

$$\text{Malate} + 2\text{Fe}^{+++} \rightarrow \text{oXalacetate} + 2\text{Fe}^{++} + 2\text{H}^{+},$$
$$2\text{Fe}^{++} + 2\text{H}^{+} + 1/2\text{O}_2 \rightarrow 2\text{Fe}^{+++} + \text{H}_2\text{O}.$$

As Winfree emphasized, the conversion of Fe^{+++} into Fe^{++} and again into Fe^{+++} is not a periodic process but a reversible reaction. However, it is a repetitive process that could induce the idea of an oscillatory reaction.

How did Belousov use the concept of biochemical cycles in his work? His list of scientific papers provides a key to his references to the cycles. It mentions an unpublished manuscript entitled "On the Importance of the Small Tricarboxylic Acid Cycle for the Natural Removal (Excretion) of Some Heavy Metals From Organisms," dated 1951. This manuscript is lost.

However, it is possible to judge Belousov's ideas in this area by the roundabout of Safronov's references to it. Safronov proposed an alternative way of excerpting heavy metals: In contrast to Belousov's idea of urinary excretion, he suggested fecal excretion. Safronov wrote that according to Belousov, urinal excretion was provided with the help of endogenic citric acid and citrates.

It is plausible that Belousov's ideas proceeded at least partially along the path of other research on this topic as represented, e.g., in the 1960 paper of Schubert and Lindenbaum, which reported on an attempt to use endogenous native citric acid and citrates for the removal of heavy metals from a human body. It is possible (with the help of fluoroacetic acid) to cause an accumulation of citric acid, which then forms chelating complexes with heavy metals. Fluoroacetic acid interferes with the tricarboxylic acid cycle and in so doing causes citric acid to accumulate in organs such as spleen and kidney, which are heavily damaged by heavy-metal ions.

Belousov also referred to the tricarboxylic acid cycle in the aforementioned manuscript, namely, "On Chemical Presumptions Concerning the Foundations of the Effective Action of the Remedies That Are Tested in the Therapy of Radiation

Damage." There, Belousov discussed the appearance of adrenaline in the blood in the course of radiation damage. According to him, adrenaline provides a number of dysfunction characteristics of radiation sickness. However, in small amounts, adrenaline stimulates the function of the tricarboxylic acid cycle and leads to the release of citric acid with the urinary excretion.

As was noted before, in Rogozkin et al. (1963), to which Belousov contributed the chemical and biochemical chapters, "respiratory ferments containing iron" were mentioned. Belousov therefore approached the behavior of iron in respiratory ferments in the course of his applied research in radiation toxicology.

Belousov probably meditated upon periodicity and oscillations in connection with the tricarboxylic acid cycle. Admittedly, Belousov combined the idea of a cycle with that of oscillations, to which he and Safronov came in their studies in analytical chemistry.

4.6   The Dialectics of Continuity and Discontinuity

This section deals with the linguistic structure of Belousov's discovery. It is neither about his formulation of what was produced and observed nor about a refinement of what was already understood. Belousov observed what he expressed in his descriptions, but his descriptions were determined by his scientific language: "Daß die Welt meine Welt ist, das zeigt sich darin, daß die Grenzen der Sprache die Grenzen meiner Welt bedeuten" (Wittgenstein 1955, p. 62).

To describe mechanical and electrical oscillations, it is enough to follow the variation of some parameter (space coordinate, electrical charge, current strength) in time. Oscillation is the repetitive variation around a central value. The value of the parameter increases, then decreases, and then increases again. As a matter of fact, one can describe Belousov's reaction by taking the concentration, say, of Ce^{3+}, as a parameter. Such a description of the BZ reaction eventually appeared in the theory of oscillations. However, Belousov did not write mathematical equations. He expressed his reaction in elementary chemical language, using a mix of chemical notions, chemical symbols, and commonsense terminology (e.g., "faster," "slowly"). He considered a number of competitive processes, observed that one of them was going slowly and latently, while the other was going rapidly and manifestly, and so on. He noticed that step by step, the former process (more precisely, a number of processes) accelerated and in the long run suddenly manifested itself as the fast and determining one.

Let me quote from Belousov's posthumously published full text here. Let us follow how he discussed the oxidation of Ce^{3+} and the subsequent reaction of bromate with bromide. The latter reaction is a network of reactions with different rates. In turn, the interaction of bromide with bromate consists of "a series of hidden, slowly moving intermediate reactions. … These are followed by several rapid concluding reactions" (Belousov 1985, p. 608). The main idea consists in the formation of free bromine in the reaction of bromide with bromate. This formation proceeds discontinuously; it noticeably starts the moment when the amount of bromide becomes sufficient and the acetonedicarboxylic acid produced in the slow oxidation of citric acid is exhausted. When the amount of acetonedicarboxylic acid is sufficient, this acid "captures" the bromine and does not allow it to appear in the reaction mixture.

According to Belousov, the appearance of free bromine results in a coloration of the entire reaction mixture. This effect is only "strengthened by the simultaneous formation of yellow tetravalent cerium ions" (Belousov 1985, p. 611).

By emphasizing the relation between continuity and discontinuity in his reaction, Belousov referred to the "well-known Landolt reaction" as a paradigmatic example. This was a staged reaction of iodate ion with sulfite described by the Swedish chemist Landolt (1831–1910; see Partington 1961, p. 13). In the Landolt reaction, free iodine is formed. However, this formation does not occur immediately upon mixture of iodate and sulfite ions "but only after a sometimes long delay, and then discontinuously" (Belousov 1985, p. 609). As noted before, Zhabotinsky showed that the release of free bromine was not essential for Belousov's reaction. However, we try to understand Belousov's course of thinking. This thinking was dressed in Russian chemistry language enriched by the dialectics of continuity and discontinuity, which can be traced back to Friedrich Engels's "Dialectics of Nature" and eventually to Hegel's "Philosophy of Nature." In fact, it can be traced back even further—to the atom–continuity controversy.

The dialectics of continuity and discontinuity was popular among Soviet scholars. More precisely, there was no other philosophy. The philosophy of science could not depart from Engels's ideas as repeated and interpreted by Lenin and Stalin.

Joseph V. Stalin's brochure *On Dialectical and Historical Materialism* (1938) cannot go unnoticed in this context.

This text became a part of the main ideological document of the Communist Party, namely A Short Course on the History of the Communist Party (1938), which purported to show "Marxism– Leninism in action." Stalin's "On Dialectical and Historical Materialism" was also included in his collected works "Voprosy Leninizma (Essays on Leninism)," which was treated as the Bible. In 1939, the 11th edition was published. It would be republished every year until Stalin's death. The Essays on Leninism were studied in the framework of "political education- metaphysics, dialectics does not regard the process of development as a simple process of growth, where quantitative changes do not lead to qualitative changes, but as a development which passes from insignificant and imperceptible changes to open, fundamental changes, to qualitative changes, a development in which the qualitative changes occur not gradually but rapidly and abruptly, taking the form of leap from one state to another. (Stalin 1972, p. 303; Russian original 1945, p. 537)

Stalin further cited Engels's statement that chemistry could be called the science of the qualitative changes of substances, occurring under the influence of changes in quantitative composition.

Stalin's writings remained extremely influential even after his death in 1953. They certainly constituted the Soviet official ideology. However, with respect to the "dialectical categories" (quality, quantity, contradiction, and the like), one should avoid oversimplification. These categories had two faces at least.

They could be invoked by careerists and fanatics, such as in the course of the 1951 discussion on quantum chemistry, viz. the chemical theory of resonance (see, e.g., Pechenkin 1995). But they also became a part of Soviet scientists' language, especially in chemistry. Thus, in his textbook on general chemistry, Boris V. Nekrasov (1955, p. 18) wrote that the main problem of chemistry was how to characterize change in the composition of substances: was substance changed continuously or discontinuously? Nekrasov's was the main textbook used in institutes for chemical education and has gone through many editions.

Many scholars took the couple "continuity–discontinuity" as a natural extension of scientific language—a kind of scientific common sense. They did not consider this couple to be determined by external forces. They "saw" a dialectics of continuity and discontinuity in the processes of phase transitions and in the periodical table of chemical elements. For them, this continuity–discontinuity connection was not a message of dialectical materialism or Stalin. Rather, continuity and discontinuity were connected in the nature of things. However, Stalin's emphasis on this couple of categories made it "actual," "important," and remembered. It may also be said that a scholar felt himself politically comfortable by referring to these categories.

Belousov was a man of science. By admitting the dialectics of continuity and discontinuity in his writings, he shared the scheme of thinking properly to Soviet intellectuals of his time.

So, three factors that determined Belousov's thinking, leading him to his reaction, were mentioned below: (1) the reflections of Belousov and Safronov on the theory of chromaticity, (2) Belousov's reflections on the reactions constituting and providing the tricarboxylic acid cycle, and (3) Soviet popular verbalism concerning continuity and discontinuity. The first two factors were generated by Belousov's applied research; the last was deeply rooted in the Soviet ideology of the postwar years. Certainly, the proposed reconstruction of this episode in the history of science is tentative and hypothetical.

4.7 Why Belousov's Paper Has Not Been Published?

Let us turn again to the discussion in first section. Th. Kuhn's philosophical conception of paradigm would help us to answer this question. Following T. Kuhn, we hold that a paradigm encompasses problem-solving activity, and since it involves typical conceptual tools, models, and shared examples, it reduces problems to "puzzles." To describe a paradigm, one must mention its: (1) symbolic generalizations, (2) ontological model, (3) values, and (4) shared examples.

As mentioned above, Winfree emphasizes that Belousov discovery did not fit into contemporary chemical theories, while Shnol points out that this discovery was not compatible with chemical thermodynamics. Winfree's and Shnol's interpretations should each be elaborated more carefully. Belousov's discovery did not contradict thermodynamics or any other existing scientific theory. Moreover, in his 1974 book Zhabotinsky explained the situation with Belousov's discovery by pointing to a confusing identification of the stationary state of a chemical system with its equilibrium state (Zhabotinsky 1974, p. 43). However, this explanation does not concern the historical reasons: The chemists, who refused to accept Belousov's discovery, could not read I. Prigogine's books on nonlinear thermodynamics and Zhabotinsky's own papers. We believe that the situation can be elucidated with the benefit of Kuhnian paradigms. Belousov's discovery did not fit into the paradigm of classical chemical thermodynamics. This paradigm can be described as follows: (1) ("symbolic generalizations") Thermodynamical functions characterize how a system is approaching its equilibrium state, (2) ("ontological model") every free system spontaneously arrives at dynamic chemical equilibrium, (3) (values) thermodynamics provides the basis for studies in chemical processes, and (4) ("shared examples") simple physical processes, say, equilibrium between a liquid and its vapor, the irreversible process of mixing of two gases.

In turn, Belousov's reaction presupposes the following: (1) Although thermodynamic functions are monotonic, the states of some systems are described by periodic functions, (2) the evolution of the system results in a stable stationary structure constituted by ordered transformations of molecules, whereas thermodynamic equilibrium, at which every system should spontaneously arrive, is provided by a chaotic set of molecular processes (approximately half of particles participate in a forward reaction, while the other half participate in a reverse reaction, and as a result, there are no changes in the system). (3) The Belousov discovery invites suspicion toward standard thermodynamics. (4) "Shared thermodynamic examples" turn out to be irrelevant.

True, this is only a part of explanation. B. P. Belousov did not belong to the scientific establishment. He was not a person of the Academy of Sciences and/or Lomonosov Moscow State University. He had not influential friends to ask for support. In addition, his noble character did not allow him to ask, to insist, and to organize the advertizing company to support his project.

In 1981, Belousov together with Zhabotinsky and some other scientists who investigated chemical oscillations and waves received the Lenin Prize. As was mentioned above, in 1981 his paper has been published.

References

Belousov BP (1959) Periodicheski deistvuyushchaya reaktsia i ee mekhanism [Periodically acting reaction and its mechanism]. In: Sbornik referatov po radiotsionnoi meditsine, 1958 (Collection of Abstracts on Radiation Medicine, 1958). Medgiz, Moscow, pp 145–147

Belousov BP (1985) A periodic reaction and its mechanism. In: Field R, Burger M (eds) Oscillations and traveling waves in chemical systems. Wiley, New York, pp 605–613

Belousov BP, Safronov AP (1959) Novye puti kachestvennogo analiza kationov v svete izbrannoi teorii tsvetnosti [New approaches to the qualitative analysis of cations from the point of view of the selected theory of chromaticity]. In: Sbornik referatov po radiotsionnoi meditsine, 1958 (Collection of Abstracts on Radiation Medicine, 1958). Medgiz, Moscow, pp 147–148 (in Russian)

Belousov BP, Gurevich ID, Mindalev LA, Nazarov VI, Skliarenko SI (1932) Neorganicheskaia khimia v prilozhenii k voenno-khimicheskomu delu (Inorganic chemistry in its application to military-chemical problems). Chemical Courses for Officers, Moscow

Coveney P, Highfield R (1995) Frontiers of complexity. Search for order in a chaotic world. Fawcett Columbine, New York, 468 p

Field R, Burger M (eds) (1985) Oscillations and traveling waves in chemical systems. Wiley, New York

Hacken H (1977) Synergetics. An introduction. Springer, Berlin

Kuhnert L, Niedersen U (eds) (1987) Selbstorganisation chemischer Strukturen. Arbeiten von F.F. Runge, R.E. Liesegang, B.P. Belousov, A.M. Zhabotinsky. Akademie Verlag, Leipzig

Nekrasov BV (1955) Kurs obshei khimii (Course of general chemistry). Moscow

Niederson U, Kuhnert L (eds) (1987) *Selbstorganisation chemischer Strukturen. Arbeiten von Friedlieb Ferdinand Runge, Raphael Eduard Liesengang, Boris Pavlovich Belousov, und Anatol Markovich Zhabotinsky.* Ostwalds Klassiker der exakten Wissenschaften. Leipzig, Bd.272

Partington JR (1961) A history of chemistry, vol 4. Martino, New York

Patterson D (ed) (1967) Pigments: An introduction to their physical chemistry. Elsevier, Amsterdam

Pechenkin A (1995) The 1949–1951 Anti-Resonance campaign in soviet science, LULL—Revista de la sociedad espanola de historia de las ciencias y de las tecnicas, vol. 18, pp 135–159

Pechenkin A (2009) On the origin of the Belousov-Zhabotinsky reaction. Biol Theor 4(2):196–215

Poleshuk V (1984) Na obshikh osnovaniakh (On common grounds). Novyi Mir 4:189–207

Prigogine I (1980) From being to becoming. Time and complexity in the physical sciences. W.H. Freeman and Co., San Francisco

Rogozkin VD, Belousov BP, Evseeva NK (1963) Radiatsionnoie deistvie tsianistykh soedinenii. Meditsinkaya kniga, Moscow

Safronov AP (1961) Apparat dlia poluavtomaticheskogo opredelenia limonnoi kisloty (An apparatus for a new method of semi-automatic determination of citric acid). Sbornik referatov po radiatsionnoi meditsine, Moscow, pp 152–153

Shnol' SE (1997) Geroi i zlodei rossiiskoi nauki. Kron-Press, Moscow

Stalin IV (1972) The essential Stalin. Major theoretical writings. Anclor Books, New York

Strogatz SH (1994) Nonlinear dynamics and chaos. Perseus Books, Reading, 467 p

Winfree AT (1984) The prehistory of the Belousov-Zhabotinskii reaction. Chem Educ 61:6661–6665

Wittgenstein L (1955) Tractatus logico-philosophicus. Jonathan, London

Zbarsky BI, Ivanov II, Mordashev CP (1951) Biologicheskaia khimia (Biological chemistry). Medgiz, Moscow

Zhabotinsky AM (1964) Periodicheskaya zhidkofaznaya reaktsia (A periodic liquid phase reaction). Izvestia Akademii Nauk SSSR 157:392

Zhabotinsky AM (1974) Kontsentratsionnye avtokolebania (Concentration self-oscillations). Nauka, Moscow

Zhabotinsky AM (1985) The early period of systematic studies of oscillations and waves in chemical systems. In: Field R, Burger M (eds) Oscillations and traveling waves in chemical systems. Wiley, New York, pp 1–5

Zhabotinsky A (1991) A history of chemical oscillations and waves. Chaos 1:379

Chapter 5
The Belousov–Zhabotinsky Reaction

Abstract Chapter 5 provides a piece of the history of the department biophysics at Physics school at Lomonosov Moscow State University. It explains why Zhabotinsky's university teacher S.E. Schnol put the problem to reproduce and explain the Belousov reaction before Zhabotinsky. It also describes how Zhabotinsky's research group had been formed. To reconstruct Zhabotinsky's way of thought we turned to the mathematical and physical ideas of Zhabotinsky's father who was a representative of the Mandelstam-Andronov school, one of the Soviet leading scientific school. Zhabotinsky's collaborator M. Korzukhin was a representative of another scientific school, the school of applied mathematics, his scientific chief was A. Molchanov, who started at the Institute of Applied Mathematics headed by M. Keldysh. Chapter 5 also describes the intellectual interaction of Zhabotinsky's research group and I. Prigogine's group which developed non-linear non-equilibrium thermodynamics. Chapter 5 is based on interviews given to the author by Zhabotinsky's colleagues and his university teachers.

Keywords Biophysics · The university department · Biological clock
Mechanism of the reaction · Self-oscillations · Applied mathematics
Differential equations · Thermodynamics

5.1 The Ideology of Biorhythms and Biological Clocks

The expansion of chemistry into biology is well-known. The rise of biochemistry and molecular biology manifested the expansion. However, by the end of the 1950s, the opposite tendency had appeared. This was the tendency to apply some fundamental biological categories in chemistry: first of all the categories of evolution and individuality. This also involved a tendency to organizational innovations. In 1958, the department of biophysics was established at the Physics Faculty of the Lomonosov Moscow State University. The department of molecular

© The Author(s) 2018   65
A. Pechenkin, *The History of Research on Chemical Periodic Processes*,
SpringerBriefs in History of Science and Technology,
https://doi.org/10.1007/978-3-319-95108-9_5

biophysics was organized at the Faculty of Molecular and Chemical Physics of the Moscow Institute of Physics and Technology in 1959 (later this chair gave life to a new Faculty (School) of this Institution, the Faculty of Physical and Chemical Biology). Shnol lectured in the course of biochemistry at the chair of biophysics; Zhabotinsky was a graduate student of that chair. It should be also noted that Zhabotinsky's first paper appeared in the journal named *Biophysics* and that in this paper he stressed the biological importance of Belousov's reaction.

To illustrate the ideological penetration of biology into chemistry, let me cite the Nobel Prize-winning chemist N. N. Semenov, who headed the Institute of Chemical Physics where Frank-Kamenetskii (see Chap. 3) was employed. Acting as the main chemical ideologist in the USSR, he repeatedly called for adoption the "experience of life science in chemistry." In particular, Semenov said the following in his 1970 address (as it is reprinted in: Semenov 1981, pp. 192–193).

In the process of its evolution, the nature created molecular machines of the highest degree of exactness and of rapidness of action, machines of extraordinary perfection…By using the principles of chemistry of living organisms, it is possible to construct a new chemistry, that is, a new control of chemical processes. This chemistry will use the principles of synthesis of the large protein molecules, and catalysts of the highest specificity will be created according to the principles of ferments …

It is not difficult to notice that this ideology can be traced back to "natural automata" in Leibniz' and Kant's writings.

However, here it should be noted that the ideology that directly formed the early context of Zhabotinsky's research was the ideology of biorhythms and biological clocks. Studies in biorhythms, that is, in daily periodic physiological changes in a living thing can be traced back to the nineteenth century. In the 1930s, biologists launched research into genetic aspects of endogenous biorhythm, that is, biorhythms that are autonomous with respect to astronomical cycles. These studies were presumably experimental. But there was a speculative element in them. Searching for foundations for biological rhythms, biologists tended to discuss the organization of living things and to presuppose an hierarchy of periodic processes in them.

One of the "classics" of the biorhythm studies was H. Bünning, who started to study periodicity in the behavior of plants in the 1930s. In 1958, he summed up his research in the book "Die physiologische Uhr" (Berlin: Springer) which gained popularity in the Soviet Union (at least Chernyavsky, Shnol, Zhabotinsky, et al. referred to Bünning's book). In 1961, an extensive proceedings of an international conference dedicated to biorhythms and biological clocks was issued. Bünning's has been translated into Russian with Shnol acting as editor. Shnol wrote in his Foreword (Shnol 1964, p. 7).

The point of the problem is to prove that the majority of living organisms have an internal, ability to measure time that is transmitted by heredity. In normal conditions the work of endogenous "biological clocks" is correlated with periodic processes in the environment.

Shnol added in a speculative manner:

What is the nature of the processes which result in a daily periodicity?... It is possible that they are based on a combination of diffusion and the processes of biosynthesis, the combination which results in periodic changes of the properties of cells. Now it is difficult to say to what extent diffusion and processes of biosynthesis can result in periodic processes with periods of short duration (a few minutes or less). Anyway it is clear that the problem of mechanism of chemical and physico-chemical periodic processes has gained important significance...

It is probable that daily periodicity is based on chemical or physico-chemical processes. Their duration is small and within a daily cycle a multitude of the "elementary chemical oscillations" takes place. Here the analogy between biological clocks and ordinary clocks comes to mind. The accuracy of mechanical clocks is determined by the stability of frequent pendulum oscillations.

As was mentioned above, in 1957 Shnol published an article concerning the oscillations of ATPase activity of actomyosin. This article was highly empirical: The author insisted that empirical facts themselves showed periodicity. As he said to the present author, a bit later he prepared an article which was never published. He allowed himself to be more speculative in the latter article. Nevertheless, the idea of fundamental chemical (or physico-chemical) periodic processes in a living thing was in the air during the late 1950s. As Shnol writes in his book (1997, pp. 156–160), by chance he learned about the content of Belousov's discovery (not about Belousov). He began to look for the person who discovered chemical oscillations and in 1958, he succeeded to become acquainted with Belousov. Belousov showed Shnol his 1951–1955 manuscript, but he refused to cooperate in research. Shnol, however, convinced Belousov to publish an abstract of his paper.

Zhabotinsky started his work in the atmosphere of biological rhythms and clocks. However, this ideology did not play a considerable role in his work, although it allowed Shnol to legitimate this work. Zhabotinsky formulated his problems and results by using the language of the physico-mathematical theory of oscillations. This ideology (the ideology of self-oscillations) will be described in the next section (see also Chap. 3, Sect. 3.4). Here let me note that material of the 1966 first All-Union (National) Conference on Oscillatory Processes in Chemistry and Biology shows that two ideologies were present at that conference (this conference was held in Pushino-na-Oke, a small city of Moscow region, where the Institute of Biophysics of the Academy of Sciences was established in 1963). A series of articles by D.S. Chernyavsky on biological periodic processes was mentioned in Sect. 5.1. This series also shows two ideologies. One reads in D.S. Chernyavsky & N.M. Chernyavskaya's article that "by nature every internal rhythm is nothing more nor less than a self-oscillatory regime of internal chemical reactions" (1960, p. 632).

A. M. ZHABOTINSKII
Summer 1983
(Photo by A. T. Winfree)

Zhabotinsky's portrait as it is published in "Oscillations and travelling waves in chemical systems"
(1985)

5.2   The Ideology of Self-oscillations

Zhabotinsky started his research by being influenced by ideology of the physico-mathematical theory of nonlinear oscillations developed by A. A. Andronov and his collaborators. The ideology of self-oscillations was the ideology of a powerful scientific community which is usually called the Mandelstam school (or the Mandelstam–Andronov school). "Self-oscillations" is a central concept of the theory of nonlinear oscillations developed by the community, the characteristic message of the theory. It formed around the concept of self-oscillations. In its early stage, the theory of nonlinear oscillations was simply the theory of self-oscillations. This concept made possible the broader application of the theory of nonlinear oscillations, whose domain was originally lumped systems, to continuous media and its subsequent progress toward synergetics (see Chap. 2, Sect. 2.6). As it was noted in the previous section, a big conference on oscillations in chemistry and biology was held in Pushino-na-Oke in 1966. Frank-Kamenetskii said in addressing this conference: "We must give a proper language to our science (he meant studies in chemical oscillations); the best way is to use language of radio-engineering because this is the most advanced field in studies of oscillations" (Frank (ed.), 1967, p. 40).

In 1965, Zhabotinsky published with his coauthor Korzuhkin an article "Mathematical modeling of chemical and ecological self-oscillatory systems." However, as Shnol recalls, he tended to use the conceptual tools of qualitative theory of differential equations from the very beginning of his research. Zhabotinsky's major book summarizing his research was published in 1974, and this book was entitled "Concentration self-oscillations." In the Foreword, he wrote that he belonged to the fourth generation of the Mandelstam's school. He regarded his father as his main teacher. His father Mark Zhabotinsky was a student of Strelkov who in turn was Mandelstam's student. In this Foreword, Shnol was only included in the general list of persons who influenced Zhabotinsky.

When the present author spoke with Shnol, I felt that something was wrong in his relations with Zhabotinsky. "Zhabotinsky is a person whose self-estimation is too-high", Shnol said. I think that a historian must be very careful by reviewing the teacher–pupil relations. However, what can be said definitely is the following: Shnol and Zhabotinsky have ideologically parted. By comparing Zhabotinsky's

"Concentration self-oscillations" and Shnol 2001 "The physical and chemical factors of biological evolution," one can notice that the former is penetrated by the physics ideology of self-oscillations and the latter keeps the ideology of biorhythm and biological clocks (Shnol connected this ideology with M. Eigen's idea of hypercycle in the book).

Zhabotinsky's work consisted of two parts: (1) He elaborated the mechanism and eventually the mathematical model of Belousov's reaction; (2) he proved that a homogeneous chemical oscillator is possible. The former part was rather paradigmatic, Zhabotinsky worked within the framework of Andronov's paradigm and his puzzle was how to find a limit cycle corresponding to the Belousov oscillator. The latter was solved by "turning around" the Bodenstein method of quasi-stationary concentration. It cannot be considered as a puzzle within the paradigm of nonlinear oscillations. Moreover, Zhabotinsky and his coauthor came close to another paradigm, the paradigm of nonlinear thermodynamics (see succeeding section). However, the ideology of self-oscillations also was in operation when Zhabotinsky proved that a homogeneous chemical oscillator is possible. The thing is that Zhabotinsky proved that the self-oscillatory homogeneous reactions are possible.

Let us turn to the former part of Zhabotinsky's research. This part was going along two lines: (1) the improvement of the Lotka model in such a way that this mathematical model would represent self-oscillations; (2) the improvement of Belousov's mechanism of his reaction in such a way that this mechanism would be self-oscillatory. These two lines had converged in the late 1960s: In 1971, the self-oscillatory model of the Belousov reaction (or more exactly, of the Belousov-type reaction) was published (Zhabotinsky et al. 1971).

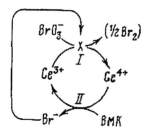

Zhabotinsky's first scheme of the Belousov reaction

5.3   Applied Mathematics

The latter part of Zhabotinsky's research (ideological with respect to the theory of nonlinear oscillations) consists in the following. Zhabotinsky and his coauthor M. Korzuhkin showed the possibility of self-oscillatory mechanism which obeyed the law of reactant masses, the basic law of chemical kinetics. In other word, they showed that self-oscillations are possible in a homogeneous chemical system.

M. Korzuhkin was a physicist according his education. When he passed through a postgraduate course (to receive Ph.D.), his supervisor was Albert Makarieviich Molchanov, a specialist in applied mathematics, who initially worked for Institute of Applied Mathematics headed by M.V. Keldysh, one of the leaders of Soviet applied mathematics.

The idea was to "turn around" the Bodenstein method of quasi-stationary concentrations. The method allowed one to make kinetics of a reaction simpler by partitioning "fast" and "slow" variations of concentrations and then by neglecting some of the variations. The mathematical scheme of the method runs as follows:

Let a mechanism of a reaction be represented by the system of differential equations

$$dx_i/dt = f_i(x, y)$$

$$\varepsilon\, dy_j/dt = g_j(x, y)$$

The former equations represent "slow motions," the latter ones represent "fast motions." If $\varepsilon \to 0$ and certain mathematical conditions are fulfilled, this system tends to the system which is simpler because its order is less, namely it tends to

$$dx_i/dt = f_i(x, y)$$

$$0 = g_j(x, y).$$

To turn this method around is meant the following: (1) To construct a simple idealized model of the chemical reaction, the model which can, however, be "non-chemical," that is, this model can contain equations that do not obey the law of reactant mass; (2) to complicate this model, to amplify it by new variables (in particular, to introduce the partition of "slow motions" and "fast motions" into the model), and to arrive at a "chemical" model consistent with the laws of chemical kinetics. This complicated model must asymptotically tend to the former model.

Zhabotinsky and Korzuhkin proved that a "non-chemical" oscillatory kinetic model can be so modified that it results in the "chemical" oscillatory model which asymptotically tends to the former.

Let me cite a simple example illuminating the above operation (Zhabotinsky and Korzuhkin 1967, p. 225). This example concerns a reaction whose idealized simple description is "non-chemical." This description is a simple differential equation: $dx_1/dt = -x_2$. The reaction is represented as a reaction of decomposition of the zero order. The "non-chemical" character of the description follows from the fact that it eventually leads to negative values of the concentration x_1.

By adding "fast motions," we transform its description into the system of equations consistent with chemical kinetics:

$$dn_1/dt = -(1/\varepsilon)\, n_1 n_0$$

$$dn_0/dt = -(1/\varepsilon) n_1 n_0 + n_2$$

Certainly, this example does not show that self-oscillatory mathematical model is consistent with the law of reactant masses. The above equations are not oscillatory at all. This example only illustrates that a very simple mathematical model with the "non-chemical" x_1 and x_2 can be transformed into a model consistent with the law of reactant masses. The latter model asymptotically tends to the original equation.

By taking the Bodenstein method into the context of the theory of differential equations, Zhabotinsky and Korzuhkin came close to H. Hacken's synergetics. These both use the "steady state condition" which can be understood as a generalization of the Bodenstein condition of quasi-stationary state (Hacken 1977, p. 9). It should be emphasized that this condition has paradigmatic significance for Hacken. For example, Hacken used the (in essence) same condition in his theory of the laser. Let us take a simple one-mode laser. After certain reasonable approximations, the theory leads to the following systems of equations

$$dn/dt = GnN - kn$$

$$dN/dt = -GnN - fN + p,$$

where n is the number of photons, N is the number of excited atoms, G is the gain coefficient for stimulated emission, k is the decay rate due to loss of photons by mirror transmission and scattering, f is the decay rate for spontaneous emission, and p is the pump strength. Suppose that N relaxes much more rapidly than n. Then we may make the quasi-stationary approximation and treat that dN/dt equals zero. By expressing N in terms of n we arrive at a first-order system for n. In Hacken's words, the behavior of N is slaved to that of n which is an order parameter.

In his *Synergetics,* Hacken treats chemical reactions, where spatio-temporal oscillations occur, with the benefit of the steady state conditions. He writes that "the onset of the occurrence of such structures is governed by principles analogous to those governing disorder-order transitions in lasers, hydrodynamics, and other systems" (Hacken 1977, p. 9).

As Hacken himself pointed out, his order parameters followed the order parameters in the L. D. Landau theory of second-order phase transitions (1977, p. 180). However, there is another source of Hacken's synergetics in the Soviet science. This is the partition of "slow" and "fast" motions in the writings of the Mandelstam–Andronov school. Originally, this operation was auxiliary: Since Andronov and Vitt's 1930 paper on relaxation self-oscillations it helped to treat "puzzles." However, step by step it became more important. The theorems of A. N. Tikhonov on differential equations with small parameter (1948, 1952) provided the necessary background to the partition. In particular, this operation culminated in Zhabotinsky's and Korzuhkin's work. Zhabotinsky and Korzuhkin used it in the paradigmatic part of their work: It allowed them to develop the Lotka equations for autocatalysis and to yield a limit cycle. Apart from this, they solved a more general problem: They theoretically showed that the concentration self-oscillations are consistent with chemical kinetics.

5.4 The I. Prigogine Nonlinear Thermodynamics

I. Prigogine (personally and in coauthorship) wrote a number of popular books. However, to understand his central concept "dissipative structure," the present author recommends to turn to his fundamental book written in coauthorship, Glansdorff-Prigogine's book "Thermodynamic theory of structure, stability, and fluctuations" (1971).

In the Introduction, Glansdorff and Prigogine formulate the nominative definition of the dissipative structure. According to Glansdorff and Prigogine (1971): "From the macroscopic point of view it is necessary to distinguish between two types of structure: equilibrium structures and dissipative structures. Equilibrium structures may be formed and maintained through reversible transformations implying no appreciable derivation from equilibrium. A crystal is a typical example of equilibrium structure. Dissipative structures have a quite different status: they are formed and maintained through the effect of exchange of energy and matter in non-equilibrium conditions. The formation of cell patterns at the onset of free convection […] is a typical example of dissipative structure" (Glansdorff and Prigogine 1971, p. 9).

Here Glansdorff and Prigogine write about the Bénard cells. In his Nobel lecture, Prigogine (1977) provided the following description of this phenomenon: It is remarkable that this new type of behavior appears already in typical situations studied in classical hydrodynamics. The example which was first analyzed from this point of view is the so-called "Bénard instability." Consider a horizontal layer of fluid between two infinite parallel planes in a constant gravitational field, and let us maintain the lower boundary at temperature T_1 and the higher boundary at temperature T_2 with $T_1 > T_2$. For a sufficiently large value of the "adverse" gradient $(T_1 - T_2)/(T_1 + T_2)$, the state of rest becomes unstable and convection starts. The entropy production is then increased as the convection provides a new mechanism of heat transport. Moreover, the state of flow, which appears beyond the instability, is a state of organization as compared to the state of rest. Indeed a macroscopic number of molecules have to move in a coherent fashion over macroscopic times to realize the flow pattern"

(Prigogine 1977). The term "entropy production" is present in the above piece. This terminology is typical for the Brussels school to which Prigogine belonged. The change of entropy can be split into two parts: the entropy production due to changes inside the system, and the flow of entropy due to interaction of it with the outside world. According to the second law of thermodynamics, the entropy production is never negative.

The nominative definition allows us to distinguish between phenomena: between equilibrium structures and dissipative structures. It does not allow to catch the specific of the phenomenon. Many phenomena which are not dissipative structures do fit this definition (say, by mentioning the dissipative structure one can refer to the earth as a whole, the earth as an open system subject to the constant flow of energy from the sun).

To formulate a *real* definition of the dissipative structure means to construct the theory of the phenomena. Prigogine's thermodynamics (this article mainly refers to Glansdorff and Prigogine 1971) are constructed according to the principle of generalizations. It has three levels of generality: regular thermodynamics of the reversible processes, linear thermodynamics of the irreversible process, and non-linear thermodynamics of the irreversible processes. Within the framework of the third level, the concept of the dissipative structure is constructed and explained.

To explain the dissipative structure, nonlinear thermodynamics formulated a number of concepts: thermodynamic flows, thermodynamic forces, entropy flow, entropy production, excess entropy. In this theory, a number of new theorems are formulated and proved.

Here we shall not reproduce these concepts and theorems. The real definition of dissipative structure cannot be formulated as a laconic sentence. We have the rather a contextual definition of the concept.

What is important for us in the context of the present book? By formulating the concept of dissipative structure, Glansdorff and Prigogine characterized only two structures as dissipative. This is the above-mentioned Benard gels and the Belousov–Zhabotinsky reaction. Only for these structures, the proper theoretical discussion has been provided in their book.

In this context, Richard Field's sentence becomes understandable. "In the mid-1960s, when study of the BZ reaction became intensive, a large body of theory of non-linear differential equations, bifurcation theory (mainly the so-called Hopf bifurcation), and already been developed in the Soviet Union (e.g. Andronov, Salnikov) and in western Europe (Hacken, Hess, Pacualt, Prigogine). Early experimental (Bray) and theoretical (Lotka) work had been done in America. However, in my opinion this body of elegant theory was going nowhere because no-well understood experimental chemical example and an experimental area where theory could be tested and extended. This led directly in my opinion, to 30 years of explosive growth and understanding of nonlinear dynamic systems. I believe that I. Prigogine's 1977 Nobel Prize (non-equilibrium thermodynamic and dissipative structures) would not been awarded, if the BZ reaction had not been discovered and its mechanism elucidated and shown to be just as predicted by theory" (R. Field's letter. 17.11.2004).

The concept of dissipative structure has been considerably expanded since the beginning of the 1970s. First of all, it was expanded over some of biological structure. There is an attempt to treat the Liesegang ring formation as a dissipative structure (Bykov 2006).

At the end of the 1960s, Prigogine and his coworkers constructed a hypothetical chemical model known as the Brusselator. The Brusselator is not a model of the BZ reaction. Prigogine and coworkers tended to "investigate the dynamic requirement for temporal oscillation, as well spontaneous pattern formation to occur in chemical systems. Both phenomena were dubbed dissipative structures by Progogine because they are supported by the dissipation of free energy. The final approach to equilibrium must be monotonic, and Prigogine's work show chemical oscillations are far-from-equilibrium phenomenon" (Field and Schellman 1999, p. 12).

True, Alan Turing should be mentioned before. "The existence of nonlinear chemical systems exhibiting temporal oscillations and even spatial formation was suggested in 1952 by Turing as chemical rationalization of morphogenesis. This work was greatly extended by Prigogine and his coworkers who in 1968 presented a simple two variable model that includes a so called trimolecular step and exhibits limit cycle behavior" (Field 2015, pp. 1530015–4).

"Trimolecular step" means an interaction of three molecular that is $2A + B$. About limit cycles, see Chap. 3, Sect. 3.4.

References

Bykov VI (2006) Modelirovanie kriticheskikh iavlenii v khimicheskoi kinetike (*Modeling of critical phenomena in chemical kinetics*). Bookseries: Synergetics, Moscow

Chernavskaya NM, Chernavsky DS (1960) Periodicheskie iavlenia v fotosinteze (The periodical phenomena in photosynthesis). Uspekhi fizicheskikh nauk 72:627–649

Field RJ (2015) Chaos in the Belousov-Zhabotinsky reaction. Mod Phys Lett 29(34):153015 (39pp)

Field R, Schellman JA (1999) Richard Macy Noyes (1919–1997), vol 77. National Academy of Sciences. Biographical Memories, Washington, DC

Frank GM (1967) Oscillatory processes in chemical and biological systems. Nauka, Moscow

Glansdorff P, Prigogine I (1971) Thermodynamic theory of structure, stability, fluctuations. Wiley-interscience, NY

Hacken H (1977) Synergetics. An introduction. Springer, Berlin

Prigogine I (1977) Noble lecture

Semenov NN (1981) Nauka i obshestvo (Science and society). Nauka, Moscow

Shnol' SE (1964) Foreword to the Russian translation of Bunning's book "Biologocal Clock". Mir, Moscow, pp 6–14

Shnol' SE (1997) Geroi i zlodei rossiiskoi nauki (Heroes and Villans of Russian Science). Moscow

Shnol' SE (2001) Geroi, konformisty i zlodei Rossiiskoi nauki. (Heroes, Conformists, and Villans of Russian Science), Moscow

Zhabotinsky AM (1985) The early period of systematic studies of oscillations and waves in chemical systems. In: Field R, Burger M (eds) Oscillations and traveling waves in chemical systems. Wiley, New York, pp 1–5

Zhabotinsky A, Korzukhin M (1967) Mathematical modeling of the kinetics of homogeneous chemical systems, Kolebatel'nye protsessy v biologicheskikh I khimicheskikh sistemakh (The oscillatory processes in chemical and biological systems). Moscow
Zhabotinsky AM, Korzukhin MD, Zaikin AM, Kreitser GP (1971) The mathematical model of self-oscillatory chemical system, Kinetika i kataliz 12:584

Chapter 6
The American Line

Abstract Chapter 6 describes W. Bray's biography and his research in chemical kinetics. Influenced by Lotka's ideas Bray conducted the oscillatory reaction which can be treated as the first homogeneous oscillatory reactions. Had the Bray research influenced Zhabotinsky's group which developed the kinetics of Belousov's reaction? Rather the Belousov-Zhabotinsky reaction created the context within which the Bray reaction has been explained and receive ample recognition. In the 1970s influenced by Prigohine's interest in the Belousov-Zhabotinsky reaction the USA specialists in chemical kinetics Richard Field, Endre Körös and Richard Noyes elaborated the mechanism of the Belousov-Zhabotinsky reaction which has been recognized as most adequate. This mechanism differs from Zhabotinsky's ideas in some points. On the base of this mechanism the mathematical model of the Belousov-Zhabotinsky reaction has been proposed. This model is named Oregonator (as differentiated from Prigogine's model–Brusselator). This chapter based on the interviews conducted by the author with S. Vavilin who cooperated with Zhabotnsky by studying the mechanism of the Belousov reaction, and with R. Field who elaborated the FKN mechanism together with Richard Noyes and Endre Körös.

Keywords Ostwald school · Lotka's equation · Hydrogen peroxide
Iodate ion · Orgonator · Mathematical model · Slow and fast motions
Scientific communications · Fellowships · Baconianism

6.1 The W. Bray Reaction

The biography of W. Bray can be found in the *Biographical Memories of the National Academy*. There, one can read the following statement about Bray's background (pp. 13–15):

"William Crowell Bray was born September 2, 1879 in Wingham, Ontario, Canada, of good English stock...

© The Author(s) 2018   77
A. Pechenkin, *The History of Research on Chemical Periodic Processes*,
SpringerBriefs in History of Science and Technology,
https://doi.org/10.1007/978-3-319-95108-9_6

His traveling fellowship took him to Leipzig, to the school of Wilhelm Ostwald, where physical chemistry had been receiving its main impulse... A long paper on the hydrogen halides in four part of the *Zeitschrift für physikalische chemie*, in 1906, and in the same journal and year, a study of the reaction of chlorine dioxide with the chlorine acids.

In Lȩipzig, he met a number of American students, including Arthur Lamb, who later achieved professional distinction. Although at that time he was very much Canadian, he joined the "American Colony Club". These contacts opened the way for an invitation in 1905 to join the remarkable group of young physical chemists..., gathered by Arthur Noyes at the Massachusetts Institute of Technology.

In 1912, he joined the group of enthusiastic young chemists gathered by Gilbert Lewis at the University of California at Berkeley".

In his book, Servos (1990, p. 163) concerning to development chemistry in the USA makes an essential remark concerning W. Bray's scientific biography. Servos pointed to the turn to chemical physics (from physical chemistry), the turn which is remarkable for American chemists influenced by Wilhelm Ostwald. This does not mean that Bray became a specialist in chemical physics like Russian chemist N. N. Semenov did. This means that Bray viewed the elementary acts of the reaction with great interest.

"W. C. Bray indeed studied at Ostwald's institute but his Ph.D. was done with Luther, an inorganic chemist. It was because of his Ostwald connection he found himself at MIT with A. A. Noyes and eventually at UC-Berkeley. He was undoubtedly an inorganic rather than a physical chemist. Henry Taube was his most well-known student, winning a Nobel Prize for his work on the rates and mechanisms of inorganic reactions. W. C. Bray's power of observation was well-known, presumably allowing him to observe the Bray Oscillations despite their very long period under his experimental conditions" (R. Field's 2/10/2017 letter to the present author).

The story of the Bray reaction has been followed by Gervellati and Greco (2017, p. 195). "Around 1916, Bray was studying the dual role of hydrogen peroxide as an oxidizing and reducing agent together with his student, Asa L. Caulkins. The reactions they studied were the oxidation of iodine to iodate ions and the reduction of iodate ions to iodine, that is

$$5H_2O_2 + I \rightarrow 2IO_3 + 2H + 4H_2O$$
$$5H_2O_2 + 2IO_3 + 2H^+ \rightarrow I_2 + 5O_2 + 6H_2O$$

both of which are involved in the IO_3^-, H^+/I_2 redox couple. This couple was selected by the two researchers because, based on thermodynamic calculations, they expected that it would catalyze the decomposition reaction of hydrogen peroxide..."

In 1916 Caulkins defended his thesis "A study of Reactions Involving Hydrogen Peroxide, Iodine and Iodate Ion" (supervised by Bray) for the degree of Master of Arts in chemistry (Cervellati and Greco, p. 195).

In turn, in 1921, Bray published his paper which later became famous: "Periodic reaction in homogeneous solution and its relation to catalysis." In this paper, he described an "interesting phenomenon" as he himself noted: at given concentration of hydrogen peroxide and iodate in the concentration range of sulfuric acid between more than 0.055 N and less than 0.110 N, oxygen development increased by periodic pulses.

The summary of that paper is cited below:

"The reactions (1) $5H_2O_2 + I_2 = 2HIO_3 + 4H_2O$, (2) $5H_2O_2 + 2HIO_2 = 5O_2 + I_2 + 6H_2O$, (3) $H_2O_2 = H_2O + 1/2O_2$ take place in solution. Reactions (1) proceeds rapidly in moderately high acid concn. And it is markedly autocatalytic. Reaction 2 proceeds slowly under most favorable conditions low H ion concn.; reaction 3 accompanies both of the others. For a soln of H_2O, KIO_3 and 0.110 N H_2SO_4, the vol. of O shows a steady increase with time, after an induction period of 7 min. For a soln similar except that the H_2SO_4 was 0.055 N the rate of evolution of O_2 was slower after 60 min than in the first expt. after 7 min. For two solns of intermediate acid concentration, 0.073 and 0.0916 N with H_2SO_4 the evolution of 0_2 was a periodic phenomenon. This is evidence in favor of the intermediate reaction theory of catalysis. This seems to be the first instance of a periodic reaction in homogeneous soln. It is solution so arranged that oxygen is evolved so slowly as not to cause bubbles that I …. could be seen to deepen and fade periodicity, so the periodicity can not be due to the periodic release of oxygen. It has also been shown that light, traces of a chloride and the presence of a suspended solid have a marked influence on the reaction."

It is remarkable that Bray turned to Lotka and Hirniak to provide a theoretical basis in his 1921 paper. Bray wrote that "the possibility of such periodicity had… been appreciated. Lotka and Hirniak independently examined the problem, each assumed a definite mechanism for the hypothetical reaction, set up differential equations for the various intermediate reactions, and a mathematical analysis set up conditions sufficient for periodicity. In Lotka's example the autocatalytic character of the second of a series of three successive reactions is an essential feature of the assumed mechanism, and it seems possible that an explanation of a present case might be found along these lines" (Bray 1921, p. 1266).

Gervellati and Greco (2017) discussed how Lotka's and Hirniak's writings influenced Bray's research.

Bray meant the second reaction in his 1910 mechanism of the reaction which is liable to show the oscillatory behavior (see above: Chap. 3, Sect. 3.2).

Bray's 1921 observation was supported by Hedges and Mayer (1926, p. 57). They included this research into the historical chain of research entitled "The periodic catalytic decomposition of hydrogen peroxide." Schemjakin and Mikhalev's book passed the Bray reaction by.

In the historical part of his paper, Stanley Farrow provided the following description of Bray's discovery. "In 1921 Bray … accidently saw the narrow area of concentrations, where the concentration of iodine increased, decreased, and then oscillated further... He observed the oscillations of iodine concentration in the solution at 25 °C and when H_2O_2 and KIO_3 are sparely distributed. He came to a big period of oscillations (about two days) and observed no oxygen evolution (homogeneous reaction)" (Oscillations and traveling waves 1985, p. 194).

Farow pointed to criticism of Bray's conclusion: the papers appeared that Bray's reaction is not homogeneous. Bray continued his research with Hermann Liebhafsky in the 1930s, and now Bray's reaction is named as the Bray–Liebhafsky reaction.

Farrow also pointed out that over fifteen years Bray's result did not attract any attention. In the 1950s, a few articles appeared where Bray's reaction is treated as heterogeneous.

William C. Bray

6.2   From the Bray Reaction to the BZ Reaction

Vavilin, who cooperated with Zhabotinsky and contributed to the BZ reaction mechanism, recalled (Vavilin 2000):

"Before Zhabotinsky (a graduate student) and me (I was a 6 year student) the problem was put to find oscillations in Bray's system by means of the photometric registration of the concentration of iodine…

By spectrophotometrically measuring the concentration of iodine and the potential of iodine–argentum electrode (the concentration of J^-), we got the classic picture of a discontinuous limit cycle for the Bray periodical reaction (when the concentration of the J_2 is taken as a slow variable, and the concentration of J^- is taken as a fast variable). Using spectrophotometric record of Ce^{4+} simultaneously with the recording of the brome–argentums electrode potential (Br^-), we saw a fairly complicated phase portraits. These results showed primarily on more complex mechanism of the Belousov reaction than the Bray reaction mechanism. They showed that the Belousov reaction required introducing more than two variables".

This is an abstract of Vavilin–Zhabotinskii's paper. "The oscillation of the I^- ion concentration during the course of the hydrogen peroxide-iodate reaction has been investigated in the soln. contg. $0.2M$ H_2O_2 plus $0.075M$ KIO_3 plus 0.1 N H_2SO_4, temp. $60°$. Results are presented in the form of the light absorption curves. On the basis of the measurements, the reaction scheme consisting of $2H^+ + 2IO_3^- + 5H_2O_2 \longrightarrow I_2 + 5O_2 + 6H_2O$, $I_2 + 5H_2O_2 \rightarrow 2H^+ + 2IO_3^- + 4H_2O$ is discussed".

Was the Bray reaction a kind of heuristic for Zhabotinsky and Vavilin? It is interesting that they used mathematics as a working tool. R. Field wrote to the present author: "I have no idea how influential this investigation was in their work on the BZ reaction. I am not at all surprised that the Bray motion in the Iodine/iodide plane is simple because iodine and iodide are closely related through the iodine-hydrolysis equilibrium. The equivalent plot in the BZ systems would be bromide ion versus bromous acid. I am sorry to have forgotten this paper" (29. 07. 2015).

It is more probable that Zhabotinsky's and his colleagues' research awakened interest in the Bray reaction. At the end of XX two reactions had been usually cited as the homogeneous oscillatory reactions: the BZ reaction and the Bray-Liebhafsky reaction. "Now that both the Belousov-Zhabotinsky and the Bray-Liebhafsky oscillators are understood, we can begin to generalize the reaction features to look for when developing similar systems" (see: Sharma and Noyes 1976, p. 4359).

6.3   The FKN Mechanism and Oregonator

Oregonator is the name of the mathematical scheme of the Belousov–Zhabotinsky reaction, the scheme which was produced Field and Noyes who together worked for University of Oregon in 1974. Oregonator followed the FKN mechanism of the

BZ reaction described by Field, Korös, and Noyes in 1970–1974 (Field et al. 1972; Field and Noyes 1974a, b). This mechanism is usually considered to be most detailed and complete (however, Zhabotinsky made some critical comments concerning this scheme—see his chapter in Field and Burger 1985).

The word "Oregonator" is a kind of copy of the word "Brusselator" under which Prigogine and his coworkers annunciated their mathematical model of the oscillating autocatalytic reaction (see: Chap. 5, Sect. 5.4).

"In order to interpret these fascinating and extremely spectacular phenomenon, G. Nicolis and I. Prigogine write, it is necessary to have the detailed mechanism of the reaction. Recently, Noyes, Field, Körös and coworkers… completed an extensive series of experiments on the malonic acid-bromate reaction and proposed a detailed kinetic mechanism comprising more than 11 steps. Fortunately, they were able to simplify their more detailed mechanism and interpret the oscillations in homogeneous solution in the term of three key substances: (a) $HBrO_2$ which seems to play the role of a switch intermediate, (b) Br^-, which seems to play the role of control intermediate, and (c) Ce^{4+}, which can be regarded as a regeneration intermediate in the sense that it is rapidly produced when the system is switched in one direction and permits thereafter the formation of the control intermediate Br^-" (Nicolis and Prigogine, p. 343).

The major part of the so-called FKN mechanism of the Belousov–Zhabotinsky reaction separates into two parts, the first (Process A) occurring at relatively high $[Br^-]$ and the second (Process B) occurring at relatively low $[Br^-]$.

Process A (High $[Br^-]$)

(1) $Br^- + BrO_3^- + 2H^+ \rightarrow HBrO_2 + HOBr$
(2) $HBrO_2 + H^+ + Br^- \rightarrow 2HOBr$

Process B (with small quantities of Br^- left), Ce^{3+} oxidized according to:

(3) $BrO_3^- + HBrO_2 + H^+ \rightarrow 2BrO_2 + H_2O$
(4) $BrO_2^- + Ce^{3+} + H^+ \rightarrow HBrO_2 + Ce^{4+}$
(5) $2HBrO_2 \rightarrow BrO_3^- + HOBr + H^+$

"The first step is rate limiting, whereas HOBr disappears quickly by combining with malonic acid. From (1) and (2), a quasistationary state is reached with a concentration

$$[HBrO_2] \approx k_1/k_2 [BrO_3^-][H^+], \quad \text{where } k_1/k_2 = 10^{-9}$$

Step 3 is rate limiting. Steps 3 and 4 taken together are equivalent to an autocatalytic generation of $HBrO_2$. A new quasistationary state is reached with

$$[HBrO_2] = k_3/2k_5 [BrO_3^-][H^+] \; *$$

where $k_3/k_5 \approx 10^{-4}$, $k_3 = 10^4 \text{ M}^{-2} \text{ s}^{-1}$.

Now from 2 and 3, it appears that Br^- and BrO_3 complete for $HBrO_2$. Autocatalytic production of the latter will be impossible as long as

$$k_2[Br-] > k_3[BrO_3^-]$$

Thus, at the critical concentration value

$$[Br-] = k_3/k_2[BrO_3^-] * *$$

the reaction switches from pathway 1–2 to pathway 3–5. As $[HBrO_2]$ increases, then from (2) Br^- is consumed, and $[Br^-]$ drops below the critical value. On the other hand, the produced Ce^{4+} regenerates Br^- according to the global reaction:

(6) $4Ce^{4+} + BrCH(COOH)_2 + H_2O + HOBr \rightarrow 2Br^- + 4Ce^{3+} + 3CO_2 + 6H^+$

Subsequently, $[Br^-]$ exceeds the threshold level ** and $[HBrO_2]$ comes back to the level given by *. In this way the occurrence of oscillations is explained qualitatively" (Nicolis and Prigogine, pp. 543–544).[1]

To appreciate the history, we recommend to read the following:

Recollections of an Observer of the Early Days of Research on Oscillating Reactions

My name is Robert Mazo. I am Professor of Chemistry Emeritus at the University of Oregon. Although I never participated in the actual research work on oscillating reactions, I was involved peripherally, and have been asked to set down my memories of that peripheral involvement. I currently live in a retirement community and consequently do not have access to a scientific library nor to any of my old documents with which to check my memory. The reader must take this into account when evaluating what is presented here; much of it is over 40 years old.

My earliest recollection of hearing about oscillating reactions comes from about 1960, when I was teaching at the California Institute of Technology. I remember a conversation with Professor Robert Scott of the University of California at Los Angeles. He told me that he had been an "outside member" on a PhD examining committee in chemical engineering; the thesis topic was something about air pollution. Among the results presented by the candidate was a measurement of the amount of some chemical species downstream from the source. This concentration oscillated. Remember, at that time oscillations in homogeneous reaction were considered a violation of the second law of thermodynamics and not worthy of consideration. Scott, however, saw no obvious problem with the experimental method, and the result worried him. He asked me if I understood it, and I replied "no." We discussed it for a little bit and, having no better explanation, conjectured that it may have been due to some hydrodynamic effect in the flowing reacting mixture. Perhaps it was. I know no more about this incident, but it has obviously stuck in my mind for over 50 years.

Now go forward to 1968. I had left Caltech and taken a position at the University of Oregon. Having been there for the requisite time, I was eligible for sabbatical leave, and took a sabbatical year in the department of Professor Ilya Prigogine at the Université Libre

---

[1]One can find authentical presentation of the FKN mechanism in Field and Burger (1985) and in Field and Schellman (1999). Here the secondary presentation in Nicolis and Prigogine (1989) is presented.

de Bruxelles. The paper of Prigogine and René Lefever on dissipative structures was new and fresh on everyone's mind; it was the subject of much discussion. Then a shot note appeared in the literature pointing out that a real oscillating chemical reaction had, in fact, been observed, what is now called the Belousov-Zhabotinsky reaction, and giving a recipe for easily reproducing it. Of course the Prigogine group immediately tried this, and interest in the subject intensified.

When I returned to Eugene in the fall of 1969, I gave a colloquium on dissipative structures to the chemistry department. I thought it was by far the most interesting thing I had learned during my leave. I discussed the Prigogine-Lefever paper, the toy mechanism they presented which they showed could result in oscillations (subsequently called the *Brusselator*), and actually demonstrated the unstirred BZ reaction in a test tube, with its characteristic stripes.

In the ensuing question period, Richard Noyes said "This is all extremely interesting, but that skeleton mechanism you presented is chemically unrealistic and can't have anything to do with the origin of the oscillations". I replied that it was not supposed to apply to the BZ reaction, but was developed to show that oscillations could occur under the normal conditions of chemical kinetics, albeit with an unrealistic, though not violating any known laws, mechanism. Noyes still didn't like it, and said, "I'm going to go into the lab and find out what really happens in that reaction!"

And so he did. He enlisted Richard Field, then a postdoc, and later E. Korös, a visiting scholar. Eventually they developed what is now known as the FKN mechanism. So Noyes did, in fact, succeed in finding out "what really happens."

But" what really happens" is quite complex. When I saw the FKN mechanism I said to Noyes, "This is so complex that only a few experts in kinetics will ever study it in detail. Yet this is an important piece of research for all physical chemists to understand. What you ought to do is boil down the mechanism to a few essential steps, ignoring all of the intermediate compounds and their reactions which are not essential for understanding the oscillation." Noyes was reluctant to consider this; he was interested in the chemistry of the individual case, not so much in the general principal. So I took my preaching on this subject to Field and had an easy time convincing him that my suggestion was worthwhile. I don't know how Field was able to convince Noyes, but the upshot was the *Oregonator*.

So, although I played no role in the actual research on oscillating reactions, in two instances I did play a role as a catalyst. I take some pleasure in even such a minor role.

There is more short description of the 1969 year situation: "Körös and Noyes working at the University of Oregon, Eugene (Noyes) became aware of the BZ reaction in the fall of 1969 from Robert M. Maso, who had recently returned from a sabbatical year in Brussels with Prigogine. They were intrigued by the BZ reaction oscillations and immediately began work to elucidate the BZ chemical oscillations as well as the origin of the observed oscillations" (Field 2015, pp. 1530015–7).

Conventionally, Zhabotinsky's approach to modeling of his and Belousov's reaction can be characterized as phenomenological (Zhabotinsky himself used this word to indicate his methodology). In contrast, the Field, Korös, and Noyes' approach can be called as chemico-physical. They concentrated on the elementary steps of the reactions.

As we have seen Zhabotinsky started by modeling Belousov's reaction as self-oscillating mechanism. "A lot of time was spent proving the homogeneity of the oscillating reaction… Work was then started in collaborating with M. D. Korzuhkin on the mathematical simulation of the Belousov reaction. Our belief was that when a suitable empirical model was obtained, it would be possible to relate the model to the actual chemical networks. … Some interesting results were obtained. We showed that replacement of the simple autocatalytic reaction in the conservative Lotka-Volterra system

$$a + b \rightarrow 2a$$

by its simplest two-elementary reaction equivalent

$$a + b \rightarrow c$$

$$c \rightarrow 2a$$

led to a four variable self-oscillatory system.

It was proved (Korzuhkin) that oscillations are possible in homogeneous chemical systems" (Field and Burger 1985, pp. 2–3).

Vavilin wrote to the present author (27, 07, 2015) that "Zhabotinsky cooperated with Korzuhkin who followed Molchanov's idea to differentiate between fast and slow processes. This method resulted in tremendous amounts of the mechanisms of the reaction and it was the dead end. I tended to apply the traditional methodology and here Zhabotinsky and me disagreed". Traditional methodology means that more importance is attached to accumulating of experimental results and to classifying the reactions.

Field, Korös, and Noyes put forward their mechanism on the base fundamental experimental research in the stile of chemical physics. Then Field and Noyes came to the mathematical model. Their line of attack on the problem is historically represented in Field and Schellman (1999). "The connection between the BZ reaction and the mathematics of non-linear dynamics was made firm by Dick and R. J. Field in 1974 by their introduction of the Oregonator, a simple model derived from the FKN mechanism and similar to Prigogine's Brusselator. … The FKN mechanism and the Oregonator were pivotal to the development of an entirely new and broadly applicable area of science".

Let me say that this differentiation of two lines of attack is conventional. Like American scientists, Zhabotinsky with coauthors used both experimental research and mathematical theoretical models. One can speak about a destination between the tendencies in research rather than between structures of research. However, it is worth to recall the word "Baconianism" coined by J. Servos in his history of American science (1986, vol. 7, p. 614). True, Servos wrote about the physical sciences in America in 1880–1930. Nevertheless by comparing two lines of attack on the mechanism of the BZ reaction one can pose the problem: could the American approach to the mechanism of Belousov-Zhabotinsky be treated in the stile of "Baconianism", whereas the Soviet (Russian) approach had been influenced by demon of mathematics?

References

Bray WC (1921) A periodic reaction in homogeneous solution and its relation to catalysis. J Am Chem Soc 43:1262

Cervellati R, Creco E (2017) Periodic reactions. The early works of W.C. Bray and A.J. Lotka. J Chem Educ 94(2):195–201. https://doi.org/10.1021/acs.jchemed.6b00342

Field RJ (2015) Chaos in the Belousov-Zhabotinsky reaction. Mod Phys Lett 29(34):153015

Field R, Burger M (eds) (1985) Oscillations and traveling waves in chemical systems. Wiley, New York

Field RJ, Noyes RM (1974a) Oscillations in chemical systems, part 4. J Chem Phys 60:1877

Field RJ, Noyes RM (1974b) Oscillations in chemical systems, part 5. J Am Chem Soc 96:2001

Field R, Schellman JA (1999) Richard Macy Noyes (1919–1997), vol 77. National Academy of Sciences. Biographical Memoirs, Washington, DC

Field RJ, Korös E, Noyes RM (1972) Oscillations in chemical systems, part II. J Am Chem Soc 94:8649

Hedges ES, Myers JE (1926) The problem of physico-chemical periodicity. Arnold & Co, London

Nicolis G, Prigogine I (1989) Exploring complexity. An introduction. Freeman, NY

Servos JW (1986) Mathematics and the physical sciences in America, 1880–1930. ISIS 77:611–629

Servos JW (1990) Physical chemistry from Ostwald to Pauling. The making of a science in America. Princeton University Press, New Jersey, 438 pp

Sharma KR, Noyes RM (1976) Oscillations in chemical systems. Part 13. A detailed molecular mechanism for the Bray-Liebhafsky reaction. J Am Chem Soc 98:4345

Vavilin VA (2000) Self-oscillations in liquid-phase chemical systems. Priroda 5:19–25

Conclusion

Until the 1960s, research in oscillatory chemical processes predominantly developed around the Liesegang phenomenon. The scholars often referred to Liesegang's experiments, as well as to Wilhelm Ostwald's explanation of these experiments or to Ostwald's research. The "shared example" was a heterogeneous chemical reaction which shows how a precipitate occurs.

The beginning of the 1960s marked the transition to a new phase, with the appreciation of the significance of Belousov's reaction for the oscillatory problematic in chemistry. In the 1970s "the FKN mechanism removed any doubt that homogeneous can do occur solely as the result of non-linear dynamic structure. The source of the instability and oscillations in the BZ reaction is made clear as resulting from negative feedback on autocatalytic process.

This led to the restructuring of research in chemical oscillations. The theory of dynamical systems has been applied to chemical oscillations. Chemical oscillations turned out to be connected with traveling waves in chemical systems, with the phenomenon of dynamic chaos.

Studies in chemical oscillations gave an impulse to the development of non-linear dynamics and stimulated its application to chemical kinetics. The chemical world now took chemical oscillations seriously, the shibboleth disappeared, and search another examples began in earnest (Field and Schellman 1999, p. 17).

Scholars who wrote about chemical periodical processes in 1960s often felt that they constructed a new research area. They had no inclination to perceive themselves as a part of the scientific movement initiated by Wilhelm Ostwald and Raphael Liesegang, and they did not cite papers written by chemists mentioned in Chap. 2 of this book, namely those authored by Kremann, Veil, Hedges, Mikhalev, and Schemjakin. Implicitly, however, they leaned upon the authority that the chemical periodicity studies gained during the first half of the twentieth century.

The scientists of the second half of the twentieth century felt the fundamental significance of the studies on chemical periodical processes and treated the general theory of oscillations as a fundamental theory; they sympathized with the idea of

© The Author(s) 2018
A. Pechenkin, *The History of Research on Chemical Periodic Processes*,
SpringerBriefs in History of Science and Technology,
https://doi.org/10.1007/978-3-319-95108-9

"oscillatory unification of science" (Andronov and Chaikin 1937, 1949; Andronov et al. 1959, 1966; Hort 1910; Barkhausen 1950). These scholars also sympathized with the Lotka's statement that "Periodic phenomena play an important role in nature, both organic and inorganic" (Lotka 1920a, p. 410). Some of them could support B. Dogadkin's thesis that "For a materialist reader, the importance of the Liesegang's rings is now clear. This picture let us to come to the conclusions that are essential for the development of our knowledge of life along with our knowledge of nature" (Dogadkin 1928, p. 58) and, perhaps, agreed with Schemjakin and Mikhalev who wrote: "The study of periodical processes is of great scientific and practical significance. We come across these processes in science, technology, and medicine" (Schemjakin and Mikhalev 1938, p. 2).

Starting with the Liesegang's research, studies in chemical oscillations were strongly connected with biology. The Belousov–Zhabotinsky reaction and its interpretation by Prigogine and his school resulted in a new direction in Biophysics and Molecular Biology. These fields started to explore a new concept, the dissipative structure.

It is remarkable that some contemporary papers can be directly traced back to the books and articles published before the World War II. One of the recent articles even[*] includes a dedication: "To the memory of R. E. Liesegang and the 75th Anniversary of the book "Physicochemical periodic processes" written by F. M. Schemjakin and P. F. Mikhalev" (Kuzmin et al. 2013, p. 363). However, one should not ignore an innovative aspect of the contemporary research. The concepts of "traveling waves," "spherical waves," and "dynamic chaos" entered the language of chemistry together with the research on the Belousov–Zhabotinsky reaction and with the discussion of this research.

Although this book did not discuss the impact of the Belousov–Zhabotinsky results on biology, specifically molecular biology and biophysics, it should be noted that there is a great special history of its impact on biology.

The oscillatory problems retain their great significance for chemistry. In 2007, Gerhard Ertl received the Nobel Prize for his studies on chemical processes on solid surfaces. His work focused on surface chemistry; he described chemical oscillations in catalytic CO oxidation reactions, namely the temporal oscillation of the rate CO_2 formation on a Pt (110) surface. During his Nobel Lecture, Ertl referred to the Prigogine's concept of dissipative structure which, as it was above shown, can be ideologically connected with the Belousov–Zhabotinsky reaction. He also referred to the Lotka–Voltera's equation and ideas, and he took them as the starting point of the mathematical modeling in his oscillatory kinetics (Ertl 2007).

Together with this history, we came back to the problems to which the second chapter is dedicated, to the problems of heterogeneous oscillatory systems and to the problems of heterogeneous catalysis. Ertl's research points to the new line in the development of problematics of periodical processes in chemistry. Nevertheless, it justifies the emphasis on the heterogeneous oscillatory processes. Ertl writes in one of his paper: "Although oscillatory kinetics in a heterogeneous chemical reaction systems had been discovered quite early namely by Fechner in 1928 in electrochemical reaction, it was only about 25 years ago that such phenomena were also

found in heterogeneous catalysis by the group of Wicke who observed rate oscillations in catalytic CO oxidation" (Imbihl and Ertl 1995, p. 697).

The present author is not a Marxist. However, he should mention that by
observing the above history, a Marxist would provide a remarkable observation: the
emergence of the new presupposes the relative recurrence of some elements of the
old.

Bibliography

Afanas'ev IB, Zeldovich IB, Todes OM (1949) O prostranstvennom raspredelenii osadkov pri kristallizatsii vzaimno diffundiruiushikh veshestv (On space distribution of precipitates during crystallization of mutually diffusing substances). Z Phys Chimii 28(2):156–179

Andronov AA, Chaikin SE (1937) Teoria kolebanii (Theory of oscillations). Gos. Izd, Moscow

Andronov AA, Chaikin SE (1949) Theory of oscillators. English language edition Solomon Lefschertz. Princeton University Press, Princeton, New Jersey

Andronov AA, Vitt AA, Chaikin SE (1959) Teoria kolebanii (Theory of oscillations). In: Zhelestzov NA (ed) 2nd edn. Fizmatlit, Moscow

Andronov AA, Vitt AA, Chaikin SE (1966) Theory of oscillations. Pergamon Press, Oxford

Annual report of the Board of Regents of the Smithsonian Institute (1857)

A Short History of Liesegang Rings. http://www.insilico.hu/liesegang/his

Barkhausen H (1950) Einführung in die Schwingungslehre, Dritte Auflage. S. Hirzel Verlag, Leipzig (1. Auflage, 1911)

Belousov BP (1958a) Periodically acting reaction and its mechanism. Sbornik referatov po radiatsionnoi medittsine (Collection of abstract on radiational medicine). Medgiz, Moscow, pp 145–148 (in Russian)

Belousov BP (1958b) Periodically acting reaction and its mechanism. In: Field R, Burger M (eds) Oscillations and traveling waves in chemical systems. A Wiley Interscience Publication, Chichester, NY (An English translation of Belousov, 1981)

Belousov BP (1981) Periodicheski deistvuyushchaya reaktsia i ee mekhanism (Periodically acting reaction and its mechanism). In: Avtovolnovye prot-sessy v sistemakh s diffuziei (Autowave processes in systems with diffusion). Gorkovskogo Universiteta, Gorkii.Izd, pp 178–186 (in Russian)

Belousov BP (1987) Eine periodische Reaktion und ihr Mechanismus. In: Kuhnert L, Niedersen U (eds) Selbstorganisation chemischer Strukturen: Arbeiten von FF Runge, RE Liesegang, BP Belousov, AM Zhabotinsky. Akademie Verlag, Leipzig, pp 73–82 (A German translation of Belousov, 1981)

Beneke K. Liesegang named in literature. http://www.uni-kiel.de/anorg/lagaly/group/klausSchiver/Liesegangnamedinliterature-2.pdf

Bray WC, Liebafsky HA (1931) Reaction involving hydrogen peroxide, iodine, and iodate ion. I Introduction. J Am Chem Soc 53:38–44

Demin NN (1961) Metabolism of acetylcholine at ionizing radiation damage. In: Problemy evoluitsii i enzimokhimii protsessov vozbuzhdenia (Problems of evolution and enzymological chemistry of excitation), pp 118–126

Dogadkin B (1928) Ob odnom liubopytnom iavlenii (About an interesting phenomenon). Publisher of Timeriazev, Academy of Agricultural Sciences, Moscow

© The Author(s) 2018

A. Pechenkin, *The History of Research on Chemical Periodic Processes*,
SpringerBriefs in History of Science and Technology,
https://doi.org/10.1007/978-3-319-95108-9

Dunin MS, Schemiakin FM (1929) Zur Frage über die Morphologie der chemischer Reaktionen im kolloiden Medium. Kolloid-Z 47(4):335–341

Ertl G (2007) Reactions at surfaces: from atom to complexity. Nobel lecture

Fechner GT (1832) Repertorium der Experimentalphysik: enthalten eine vollständige Zusammenstellung der neuen Fortschritte dieser Wissenschaft: als Supplement zu neuern Lehr- und Wörterbüchen der Physik. Bd.1-3. Verlag von Leopold Vos, Leipzig

Field RJ, Korös E, Noyes RM (1972) Oscillations in chemical systems II. Thorough analysis of temporal oscillations in the bromate-cerium-malonic acid system. J Am Chem Soc 94:8649–8664

Field R, Schellman JA (1999) Richard Macy Noyes (1919–1997). National Academy of Sciences. Biographical Memoirs, vol 77. Washington, DC

Gervardt YG, Frank-Kamenetskii DA (1942) Kolebania kholodnogo plameni (Oscillations of the cold flame), Department of Chemical Science, Izvestia Akademii nauk AN SSSR, pp 210–215

Henisch HK (1970) Crystal Growth in Gels. The Pennsylvania State University Press, London

Hildeband JH, Bray WC, National Academy Biographical memoirs, vol 36, pp 13–24

Hirniak J (1911) Zur Frage der periodischen Reaktionen. Zeitschrift für physkalische Chemie. Bd. 75:675–680

Hort W (1910) Technische Schwingungslehre, 2 Aufl. Julius Springer, Berlin

Imbihl R, Ertl G (1995) Oscillatory kinetics in heterogeneous catalysis. Chem Rev 95:697–733

Kolebania i begushie volny v khimicheskikh sistemakh (1988) (trans: Zhabotinsky A (ed)). Originally published as Field R, Burger M (ed) Oscillations and traveling waves in chemical systems (1985)

Kränzlein G (1935) Zum 100 jährigen Gedächtnis der Arbeiten von F. F. Runge. Angew Chem 48(1):1–3. https://doi.org/10.1002/ange.19350480103

Kuzmin VI, Gadzanov AF, Tytik DL, Busev SA, Revina AA, Vysotskii VV (2013) Kinetika i formirovanie koletz Lieseganga ('Kinetics and formation of Liesegang rings'). J Struct Chem 54(2):363–378

Lichtwitz L, Liesigang RE, Spiro K (1935) Medizinische Kolloidlehre; Physiologie, Pathologie und Therapie in kolloidchemischer Betrachtung. T. Steinkopff, Dresden, XII+ 1084

Liesegang RE (1913) Geologische diffusion. Leipzig Verlag von Theodor Steinkopff, Dresden, Mit 44 Abbildungen. Reprinted: BiblioBazaar (2009)

Liesegang RE (1915) Die Achate. Dresden und Leipzig Verlag von Theodor Steinkopff

Liesegang R (1926a) Kolloidlehre, 2 Auflage. Naturwissenschaftliche Reihe. Einführung in einfachsten. Dresden und Leipzig Verlag von Theodor Steintopf

Liesegang R (1926b) Kolloidlehre. Einführung in einfachsten. Versuchen Natura, Verlag Bundigung. Erster Teil. Republished in 1951.11

Lotka (1910) Zur Theorie der periodischen Reaktionen. Zeitschrift für physikalische Chemie 72:508–511

Lotka A (1920a) Analytical note on certain rhythmic relations, organic systems. Proc Natl Acad Sci USA 6:410–414 (American Chemical Society 42, p 1595)

Lotka A (1920b) Undamped oscillations derived from the law of mass action. J Am Chem Soc 42:1595

Mikhail Semenovich Dunin (1986) All Union Academy of Agricultural Sciences (VASKHNIL)

Mikhalev P, Schemjakin FM (1933) Emissionno-volnovaia teoria (The emission-wave theory). Zhurnal Obshei Chimii 3:1001–1004

Munk af Rosenschöld PS (1934) Regelmäßig intermittierendes Leuchten des Phosphors, Ann Phys Chem 32, Leipzig 1834. Poggendorf JC (ed) Dissertation nr. XVII, pp 216–217

Nicolis G, Prigogine I (1977) Self-organization in nonequilibrium systems, vol XII. Wiley, New York, 491 pp

Ostwald W (1896–1902) Lehrbuch der allgemeinen Chemie. In: zwei Bänden. Zweite, umgearbeitete Auflage. Zweiten Bandes Zweiter Teil. Verwandtschaftslehre. Erster Teil. Verlag von Wilhelm Engelmann, Leipzig

Ostwald W (1896) Elektrochemie. Ihre Geschichte und Lehre. Verlag von Veit & Come, Leipzig, 1182 S

Ostwald W (2003) Lebenslinien – Eine Selbstbiographie. Zweiter Teil. Leipzig. 1887–1905. Verlag der Sächsischen, Stuttgart, Leipzig

Pechenkin A (2002) The concept of self-oscillations and the rise of synergetic ideas in the theory of nonlinear oscillations. Stud Hist Philos Mod Phys 33:269–295

Pechenkin A (2015) Does Prigogine's non-linear thermodynamics support popular philosophical discussions of self-organization? Acta Baltica Historiae et Philosophiae scientiarum 3(2)

Pechenkin A (2016) The story of a book. Almagest 7(2):22–49

Quincke G (1902) Über unsichtbare Flüssigkeitsschichten und die Oberflächenspannung flüssiger Niederschläge bei Niederschlagmembranen, Zellen, Kolloiden und Gallerten. Annalen der Physik 7:631–744

Quincke GH (1924) Nature 113:280–281

Ramaiah KS (1939) On Liesegang rings. Proc Indian Acad Sci Math Sci 9(6):467–478

Salnikov IE (1948) On the theory of periodic homogeneous reactions. Doklady AN SSSR 60:405

Safronov AP (1959) O mekhanizme udalenia iz organizma inkorporirovannogo polonia (On mechanism of exertion of the incorporated polonium from the organism). Sbornik referatov po radiatsionnoi medithine. Moscow, pp 96–97

Schemjakin FM (1930) Über die Morphologie chemischer Reaktionen in kolloiden Medien. Mitteilung II. Kolloid-Zeitschrift 50(4):58–65

Schemiakin FM (1932) Über die Morphologie chemischer Reaktionen in kolloiden Medien. Mitteilung III. Kolloid-Zeitschr LVIII. Heft 3 (1932), p 325

Schemjakin FM (1939) Konspekt lektsii po fizicheskoi khimii (Synopsis of lectures on psycho-chemistry). Part 1. The Military Academy of Mechanization and Motorization of Red Army, Moscow

Schemjakin FM, Witt AA (1935) On the theory of physico-chemical periodical processes. Acta Physiochimica U.R.S.S. 121(2):171–175

Schemjakin FM, Karpov AN, Brusentsov AN (1965) Analiticheskaia khimia (Analytical chemistry. For pharmaceutical higher schools and faculties). Vyshaia Shkola, Moscow

Schleußer CA (1922) Diffusionsvorgange in Gelatine. Kolloid-Z 31:347–352

Shnol' S (1958) On spontaneous transitions of the medications of actomyosin from one state to another state. Voprosy meditsinskoi khimii [Essays on Med Chem] 4:433–440

Shnol' SE (2010) Geroi, zlodei, konformisty rossiiskoi nauki. Librokom, Moscow

Stalin IV (1945) O dialekticheskom i istoricheskom materializme, Voprosy Leninizma. OGIZ, Moscow, pp 535–563

Stäudel L, Wöhrmann H (1998) Runge-Bilder aus der Experimentierkiste. Praxis. Schule 5–10, 9. Heft 1, S.24–28

Stern KH (1967) Bibliography of Liesegang rings. The second edition. United States Department of Commerce. National Bureau of Standards. Miscellaneous Publication, 292 pp

Sukharev YI, Markov BA, Krupnova TG (2005) Energy laws of the structuring of gel phase. Chem Bioecol Izvestia Cheliabenskogo nauchnogo tsenta 4(30):120–128

Vavilin V. Self-fluctuation of iodide ion concentration during the iodate-catalyzed decomposition of hydrogen peroxide. Zhurnal Fizicheskoi Khimii 44(5):1345–1346

Vavilin VA, Zhabotinskii AM (1967) Oscillatory reaction of decomposition of hydrogen peroxide. Kolebatelnye prozessy v khimicheskikh i biologicheskikh sistemakh. Nauka, Moscow, p 220

Vitt AA, Schemjakin FM (1935) O teorii fiziko-khimicheskikh periodicheskikh protsessov (On the theory of physico-chemical periodical processes). Zhurnal Obshchei Chim 5(6):814–817

Volter BV (1985) Kto otkryl kolebatelnye khimicheskie reaktsii (Who discovered the oscillatory chemical reactions?) Chimia i zhizn' 2:8

Volter B (1988) 'Legenda and byl' o khimicheskikh kolebaniakh (the chemical oscillations: myth and reality). Znanie-sila 4:33–37. See also S. P. Kurdiumov' sait. http://spkurdyumov.ru/introduction/legenda-i-byl/ (in Russian)

Whittaker E. A history of the theories of ether and electricity. The classical theories. Thomas Nelson and Sons Ltd, London